Theory
of
Electrical Filters

Theory of Electrical Filters

J. D. Rhodes

*Department of Electrical and Electronic Engineering,
University of Leeds*

A Wiley—Interscience Publication

JOHN WILEY & SONS

London · New York · Sydney · Toronto

Copyright © 1976, by John Wiley & Sons Ltd.

All right reserved.

No part of this book may be reproduced by any means, nor transmitted, nor translated into a machine language without the written permission of the publisher.

Library of Congress Cataloging in Publication Data

Rhodes, John David.
    Theory of electrical filters.

    'A Wiley—Interscience publication.'
    1. Electric filters. 2. Electric networks.
I. Title.
TK7872.F5R48   621.3815'32   75-30767

ISBN 0 471 71806 8

Typeset in IBM Century by Preface Ltd, Salisbury, Wilts and printed in Great Britain by The Pitman Press Ltd, Bath

Preface

Originally, circuit theory developed from the analysis of the interconnection of discrete lumped components. This naturally led to the external characterization of a network in terms of its portal behaviour. Since that time numerous techniques have been established for the synthesis of linear, time-invariant networks from given external characterizing functions and have included distributed and digital networks in addition to the lumped case.

Several books are available which cover these aspects of network theory. Most have a chapter or section covering the approximation problem which deals with the construction of certain classes of external characterizing functions from prescribed specifications. In the frequency selective filter case certain examples are well known where the closed-form solution to the synthesis problem has been obtained for specific solutions to the approximation problem. Forty years ago this was achieved for maximally flat filters, twenty years ago for Chebyshev filters and more recently for elliptic function filters with the resulting explicit formulas for element values playing a significant role in the application to filter design. In a sense, one may feel that these results represent the ultimate objective of any subject area in bringing the design problem to a level at which non-specialists may readily use the results. Furthermore, for the specialist, these results establish the close relationship between the approximation and synthesis problems which should not necessarily be treated as separate entities.

The purpose of this book is twofold. In the case of the specialist, a complete rigorous treatment is given for the more meaningful solutions to the approximation problem for lumped, distributed and digital prototype filters on which most frequency selective devices can be modelled. The synthesis problem for these particular networks is tackled and explicit formulas or basic design algorithms derived. For the non-specialist, the explicit design formulas may be used directly without a complete understanding of the methods used to obtain them, thus providing powerful tools for the practising engineer.

Several original results are given in this book on the approximation problem. These have been developed in an attempt to provide a complete coverage of the subject as applied to lumped, distributed and digital filters. These mainly occur in the distributed and consequently the digital cases. Thus this book is fundamentally a research text.

The material has been strictly limited to analytical solutions for the frequency selective filter problem. This forms the foundation on which related problems such as broad-band matching, directional filtering, etc., may be developed and also provides an insight into the approach which may be used for numerical solutions to constrained filter problems for which analytical techniques are not available.

Finally, I should like to thank all those students who have participated in the research programme which has led to some of the original material presented in this book.

Contents

1 The Approximation Problem 1
   1.1 Introduction 1
   1.2 Minimum Phase Transfer Functions 6
   1.3 Equiripple Response Characteristics 12

2 Amplitude Approximations for Lumped Networks 18
   2.1 Introduction 18
   2.2 Maximally Flat Response 20
   2.3 Chebyshev Response 22
   2.4 Inverse Chebyshev Response 28
   2.5 Elliptic Function Response 30
   2.6 Synthesis of Ladder Networks 35
   2.7 Explicit Formulas for Element Values in Chebyshev
       Filters 40
   2.8 Summary of Results for Chebyshev and Maximally Flat
       Filters 47
   2.9 Explicit Formulas for Element Values in Elliptic
       Function Filters 50
   2.10 Summary of Results for Elliptic Function and Inverse
        Chebyshev Filters 59
   2.11 Determination of the Degree of the Prototype Filter . . 64
   2.12 Frequency Transformations and Impedance Scaling . . 68
   2.13 Approximate Design Techniques for Band-stop and
        Band-pass Filters 71

3 Phase Approximations for Lumped Networks 76
   3.1 Introduction 76
   3.2 Maximally Flat Linear Phase Polynomial 77
   3.3 Equidistant Linear Phase Polynomial 83
   3.4 Equidistant Constant Phase Delay Polynomial . . . 88
   3.5 Arbitrary Phase Polynomials 90
   3.6 Maximally Flat Logarithmic Phase Polynomial . . . 95
   3.7 All-pass Networks and Reflection Filters 101

4 Simultaneous Amplitude and Phase Approximations for Lumped Networks ... 105
- 4.1 Introduction ... 105
- 4.2 Constant Amplitude Filters with Phase Equalization ... 106
- 4.3 Linear Phase Filters with Amplitude Equalization ... 109
- 4.4 Optimum Maximally Flat Constant Amplitude and Linear Phase Response ... 110
- 4.5 Optimum Maximally Flat Constant Amplitude and Logarithmic Phase Response ... 118
- 4.6 Finite-band Approximations to Constant Amplitude and Arbitrary Phase Response ... 121
- 4.7 Prototype Synthesis Procedure for Transmission Type Filters ... 127

5 Amplitude Approximations for Distributed Networks ... 134
- 5.1 Introduction ... 134
- 5.2 Stepped Impedance Transmission Line Filters with Maximally Flat and Chebyshev Response Characteristics ... 136
- 5.3 Explicit Formulas for Element Values in Chebyshev Stepped Impedance Transmission Line Filters ... 139
- 5.4 Summary of Results for Chebyshev and Maximally Flat Distributed Prototype Filters ... 146
- 5.5 Interdigital Filters with Maximally Flat and Chebyshev Response Characteristics ... 149
- 5.6 Explicit Formulas for Element Values in Chebyshev Interdigital Filters ... 151
- 5.7 Summary of Results for Chebyshev and Maximally Flat Interdigital Filters ... 154
- 5.8 Fourier Coefficient Design Technique for Stepped Impedance Transmission Line Filters ... 157
- 5.9 Explicit Formulas for Element Values in Arbitrary Stepped Impedance Transmission Line Filters ... 166

6 Phase Approximations for Distributed Networks ... 171
- 6.1 Introduction ... 171
- 6.2 Exact Linear Phase Polynomials ... 172
- 6.3 Maximally Flat Distributed Linear Phase Polynomial ... 172
- 6.4 Equidistant and Arbitrary Distributed Linear Phase Polynomials ... 178
- 6.5 Distributed All-pass and Reflection Filters ... 180

7 Simultaneous Amplitude and Phase Approximations for Distributed Networks ... 184
- 7.1 Introduction ... 184

7.2	Constant Amplitude Filters with Exact Linear Phase	185
7.3	Constant Amplitude Filters with Phase Equalization	188
7.4	Linear Phase Filters with Amplitude Equalization	188
7.5	Optimum Constant Amplitude and Linear Phase Filters	190
7.6	Synthesis of Generalized Interdigital Filters	193

8 Digital Filters 206
8.1 Introduction 206
8.2 Basic Digital Wave Filters 207
8.3 Selective Linear Phase Filters 210

Appendix: Miscellaneous Amplitude Approximations 212
A.1 Generalized Chebyshev Functions with Prescribed Poles . 212
A.2 Generalized Chebyshev Functions with Prescribed Zeros . 214
A.3 Even Polynomials and Functions Equiripple over Two Bands 216
A.4 Maximally Flat Odd Polynomial Approximating a Constant 216
A.5 Maximally Flat Odd Function Approximating a Constant 217
A.6 Equiripple Odd Function Approximating a Constant . . 217
A.7 Equiripple Two-band Odd Polynomial Approximating Zero 218

References 219

Index 223

CHAPTER 1

The Approximation Problem

1.1 INTRODUCTION

Passive network theory is probably one of the most well-developed areas of electronic engineering. Apart from the compact analysis procedures, both simple and sophisticated synthesis methods are available to realize networks with prescribed functional behaviour at external ports. For lumped networks, the characterizing functions are finite and rational in the lumped complex frequency variable p, whereas for distributed or digital networks, which are restricted to be commensurate,* the complex frequency variable becomes $t = \tanh \tau p$ or $z = e^{-2\tau p}$, resulting in a periodic steady-state frequency response.

In the case of lumped networks, for example, the input impedance $Z(p)$ for a one-port network (Figure 1.1.1) may be shown by analysis to be a p.r.f. (positive real function) i.e.

$$Z(p) = \text{real for } p \text{ real}$$
$$\text{Re } Z(p) > 0 \text{ for Re } p > 0 \quad (1.1.1)$$

By utilizing one of the available synthesis procedures, e.g. Brune, Bott and Duffin, Darlington, etc., the positive real condition may also be shown to be sufficient, e.g.[1.1], i.e. $Z(p)$ may always be realized as the input impedance of a specific interconnection of resistors, capacitors, inductors and transformers with non-negative element values.

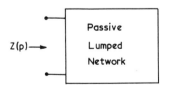

Figure 1.1.1   One-port network

---

*A commensurate distributed network is one which may be decomposed into an arbitrary connection of homogenous coupled lines, or the equivalent, of the same electrical length. In the digital case the input signal is a sequence formed from periodically sampling an analogue signal, or the equivalent, and the network possesses synchronized delay elements.

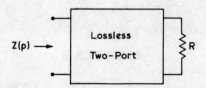

Figure 1.1.2   Darlington realization

In the case of the Darlington realization, $Z(p)$ is realized as the input impedance of a lossless reciprocal two-port network terminated in a non-negative resistor as shown in Figure 1.1.2.

The significance of this form of realization becomes apparent when consideration is given to exciting a network from a generator with a resistive internal impedance. For convenience, and without loss of generality due to the ability to scale impedance values, we shall assume that the generator impedance is 1Ω. Furthermore, we shall characterize the reciprocal lossless two-port by a scattering matrix

$$[S(p)] = \begin{bmatrix} S_{11}(p) & S_{12}(p) \\ S_{12}(p) & S_{22}(p) \end{bmatrix} \tag{1.1.2}$$

which is normalized to the input port impedance of 1Ω and the output port impedance $R\Omega$ as shown in Figure 1.1.3. Since the two-port coupling network between generator and load is lossless and passive, it may be shown that $[S(p)]$ is analytic in $\operatorname{Re} p \geqslant 0$ and $[S(p)]$ satisfies the unitary condition

$$[S(j\omega)][S^+(j\omega)] = [1] \tag{1.1.3}$$

where $[1]$ is the unity matrix and $+$ denotes taking the complex conjugate of the transpose matrix.

The reflection coefficient or scattering reflection coefficient $S_{11}(p)$ is related to $Z(p)$ through

$$Z(p) = \frac{1 + S_{11}(p)}{1 - S_{11}(p)} \tag{1.1.4}$$

or

$$S_{11}(p) = \frac{Z(p) - 1}{Z(p) + 1} \tag{1.1.5}$$

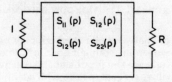

Figure 1.1.3   Normalized lossless two-port

and if $Z(p)$ is a p.r.f., then $S_{11}(p)$ is a b.r.f. (bounded real function) satisfying the conditions,

$S_{11}(p)$ is real for p real
$|S_{11}(p)| < 1$ for Re $p > 0$
(1.1.6)

The second condition may be replaced by

$S_{11}(p)$ analytic in Re $p > 0$
$|S_{11}(j\omega)| \leqslant 1$
(1.1.7)

From (1.1.3) we also have

$$|S_{11}(j\omega)|^2 + |S_{12}(j\omega)|^2 = 1 \qquad (1.1.8)$$

which implies that in addition to the transfer scattering coefficient $S_{12}(p)$ being analytic in Re $p > 0$,

$$|S_{12}(j\omega)| \leqslant 1 \qquad (1.1.9)$$

Thus, given $S_{12}(p)$ as a bounded real function, an $S_{11}(p)$ satisfying (1.1.8) may be constructed, although not necessarily uniquely, which is bounded real, and from which $Z(p)$ may be determined and consequently the network synthesized.

$S_{12}(p)$ determines the transmission properties of the signal from the generator to load. In this book, we shall be considering, in the main, the application of steady-state signals and consequently the behaviour of $S_{12}(p)$ along $p = j\omega$, or the properties of the amplitude $A(\omega)$ and phase response $\psi(\omega)$ given by

$$S_{12}(j\omega) = A(\omega) e^{j\psi(\omega)} \qquad (1.1.10)$$

Since $S(p)$ is a finite rational function in p for lumped networks or t for distributed* and digital networks, $A^2(\omega)$ is also a finite rational function in ω or $\tan \tau\omega$ and $\psi(\omega)$ is the inverse tangent of a rational function.

The ideal low-pass filter response is shown in Figure 1.1.4. For an input signal in the time domain with a band-limited spectrum contained within the passband, then the output signal is merely delayed by a constant time displacement and no phase or amplitude distortion occurs. Signals containing frequencies outside the passband are completely rejected. For distributed and digital filters, additional periodic passbands occur.

This book is entirely concerned with obtaining realizable rational transfer functions of minimum degree with respect to a given specification which produce response characteristics approximating to

---

*Occasionally, factors of the form $\sqrt{1-t^2}$ occur but may readily be accommodated either directly or by the introduction of a variable $t' = \tanh \frac{1}{2}\tau p$ which guarantees rational functions.

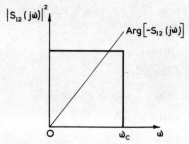

Figure 1.1.4 Ideal low-pass filter response

the required amplitude response in both the passband and stopband, and the phase response in the passband. This is the approximation problem. Lumped and distributed filter prototypes are considered, allowing most frequency selective devices which are small or comparable in size to the wavelength of operation respectively to be designed. Additionally, digital filters, as a theoretical extension of the distributed case, are included due to the increasing demand for sophisticated filtering on number sequences. In an attempt to keep the material reasonably compact, only work which possesses an analytical solution at the present time is presented. It is believed that a good understanding in these analytical techniques will assist when formulating the numerical solution in any particular instance in which an analytical solution is not known.

The problems of amplitude and phase approximation are considered independently for both lumped and distributed networks before the combined amplitude and phase approximation is developed. It is well known that the solution to the amplitude problem independent of any phase considerations is important in voice communication, since the human ear is relatively insensitive to this phase distortion. Other special signals also fall into this category in addition to ones which can tolerate some amplitude distortion with a close approximation to linear phase. However, for high capacity communications systems, selective linear phase filters are important, thus requiring solutions to the approximation problem for combined amplitude and phase constraints. The differences in this case between lumped distributed and digital solutions are highlighted and restricted forms of transfer functions relevant to each case are treated.

Once the solution to the approximation problem has been obtained, then classical solutions to synthesis problems could be applied. However, in most of the cases where analytical solutions do exist for the approximation problem closed-form expressions occur for the reflection coefficient of the network. From this, one may formally tackle the synthesis problem in an attempt to obtain explicit formulas for the element values within the filter, thus reducing the filter design problem to the use of a set of formulas without the need to use sophisticated numerical synthesis programmes.

For lumped prototype networks, explicit design formulas are given for maximally flat, Chebyshev, inverse Chebyshev and elliptic function filters covering all of the more important solutions to the amplitude approximation problem. For the basic distributed prototype network consisting of cascaded transmission lines, explicit formulas are obtained for the maximally flat and Chebyshev cases which are then extended to the important, in an engineering sense, interdigital filter. These results are also modified for the digital wave filter exhibiting low sensitivity with regard to multiplier values.

For phase and combined amplitude and phase solutions to the approximation problem, in several cases explicit formulas for element values are obtained for the associated reflection filters. For transmission-type filter realizations, although explicit formulas do not exist, simple design algorithms are given which minimize the task of determining the element values. In the distributed domain, the important class of generalized interdigital filters are considered in detail.

Since explicit design formulas or simple design algorithms are obtained for the prototype filters, it is of course possible to use these results for any filter design without a detailed knowledge of the solution to the approximation problem and the processes used to obtain these results. Thus, the purpose of this book is twofold. It provides a detailed treatment of the approximation problem and the synthesis processes necessary to obtain the design formulas, for those who wish to study these aspects in detail and possibly extend the results to more demanding requirements. Secondly, practising design engineers could use the design formulas directly without a basic understanding of the methods used to obtain them. In an attempt to assist this latter category, summaries of design formulas are given at appropriate places in the text.

In the remaining sections of this chapter, general points are considered with regard to the approximation problem. Limitations on the use of minimum phase functions are shown to suggest the necessity for non-minimum phase transfer functions for combined solutions to the amplitude and phase approximation problem. The optimum nature of equiripple solutions to the amplitude response is established for finite band approximations recovering the maximally flat solution as the limiting case. Equiripple solutions to the constant group delay problem are shown to be non-optimum although the corresponding solution to a linear phase response is optimum. Additionally, in an attempt to obtain analytical solutions, the importance of interpolating to a desired phase response is established.

Chapters 2, 3 and 4 present material on lumped prototype networks related to amplitude or phase or combined amplitude and phase approximations. Chapter 5 includes material of an advanced nature on the amplitude approximation for distributed filters which have no direct lumped counterparts. Chapter 6 establishes the difference between phase approximation in the distributed domain as compared with the lumped case and leads in to Chapter 7, which covers combined amplitude and phase approximations for distributed filters with particular reference to the realization in the form of generalized interdigital networks. In Chapter 8, the minor differences between digital and distributed solutions to the approximation problem are introduced and relevant additional material presented. Finally, in the Appendix, certain analytical solutions to the approximation problem which exist and may be important in certain applications are summarized.

1.2 MINIMUM PHASE TRANSFER FUNCTIONS

The transfer function of a stable minimum phase system is defined as one in which there are no poles or zeros in $\operatorname{Re} p > 0$, that is,

$$S_{12}(p) = \frac{N(p)}{D(p)} \tag{1.2.1}$$

and $N(p)$ and $D(p)$ are Hurwitz polynomials

$$N(p) \neq 0, \qquad D(p) \neq 0 \qquad \operatorname{Re} p > 0 \tag{1.2.2}$$

The terminology minimum phase arises from the fact that under the constraint (1.2.2) the phase shift of $\operatorname{Arg} S_{12}(j\omega)$ is a minimum over the range $-\infty \leqslant \omega \leqslant \infty$ for a given $|S_{12}(j\omega)|^2$.

Any finite linear physical system which is stable where energy may only travel from the input to output port along a single path, is normally minimum phase. Examples are ladder networks, waveguides, cables, etc. Although the above statement may be impossible to prove in general it may readily be demonstrated for systems or networks which are locally stable and may be represented as a ladder decomposition shown in Figure 1.2.1.

If the network is locally stable $Z_r(p)$ must be devoid of poles and zeros in $\operatorname{Re} p > 0$, although $\operatorname{Re} Z(j\omega)$ could be negative in an active region. Since the overall network is stable, $D(p)$ is Hurwitz and, since $S_{12}(p)$ can only be zero when $Z_r(p)$ is zero or infinite isolating the input from output, $N(p)$ must also be Hurwitz.

With minimum phase systems or networks, there is a unique relationship between the amplitude and phase response at real frequencies. This relationship is obtained by considering the functions

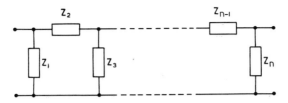

Figure 1.2.1 Ladder representation of minimum phase networks

$$\frac{\ln[S_{12}(p)]}{p^2 + \omega_0^2} \quad \text{and} \quad \frac{\ln[S_{12}(p)]}{p(p^2 + \omega_0^2)} \qquad (1.2.3)$$

which are analytic functions in Re $p > 0$. By performing a contour integration enclosing the right half-plane it may be shown that,[1,2]

$$-\psi(\omega_0) = \frac{\omega_0}{\pi} \int_{-\infty}^{\infty} \frac{\alpha(\omega)}{\omega^2 - \omega_0^2} d\omega \qquad (1.2.4)$$

and

$$\alpha(\omega_0) = \alpha(0) + \frac{\omega_0^2}{\pi} \int_{-\infty}^{\infty} \frac{\psi(\omega)}{\omega(\omega^2 - \omega_0^2)} d\omega \qquad (1.2.5)$$

where

$$S_{12}(j\omega) = e^{-\alpha(\omega)+j\psi(\omega)} \qquad (1.2.6)$$

forming a pair of Hilbert transforms.

For distributed and digital filters where $S_{12}(t)$ is a rational function in $t = \tanh \tau p$, the direct relationship emerges in the form of the Wiener–Lee transform as[1,2]

$$\alpha(\omega) = a_0 + \sum_{r=1}^{\infty} a_r \cos 2r\tau\omega \qquad (1.2.7)$$

and

$$-\psi(\omega) = \sum_{r=1}^{\infty} a_r \sin 2r\tau\omega \qquad (1.2.8)$$

To consider the phase response when minimum phase filters approximate to ideal amplitude characteristics we shall consider two examples. Let

$$|S_{12}(j\omega)| = \begin{matrix} 1 & |\omega| < 1 \\ A & |\omega| > 1 \end{matrix} \qquad (1.2.9)$$

where $A < 1$. From (1.2.4) and (1.2.6), since $\alpha(\omega)$ is an even function, we have

$$-\psi(\omega) = \frac{-2\omega}{\pi} \int_1^\infty \frac{\ln A}{y^2 - \omega^2} dy \qquad (1.2.10)$$

$$= \frac{-\ln A}{\pi} \int_1^\infty \left(\frac{1}{y - \omega} - \frac{1}{y + \omega} \right) dy$$

or

$$\psi(\omega) = \frac{-\ln A}{\pi} \ln \left| \frac{1 - \omega}{1 + \omega} \right| \qquad (1.2.11)$$

The associated group delay T_g is given by

$$T_g = \frac{-d\psi(\omega)}{d\omega} = \frac{-2 \ln A}{\pi} \frac{1}{|1 - \omega^2|} \qquad (1.2.12)$$

which increases to infinity at the cut-off frequency $|\omega| = 1$ with $1/T_g$ possessing a quadratic behaviour in the passband as illustrated in Figure 1.2.2. This characteristic is a piecewise linear approximation to the two-band approximation problem where a minimum level of attenuation is approximated over the entire stopband similar to the optimum elliptic function filter presented in Chapter 2. However, for maximally flat or Chebyshev filters, the behaviour in the stopband is maximally flat around $\omega = \infty$. Thus, for this class of filters, an approximate piecewise linear representation is

$$|S_{12}(j\omega)| = \begin{array}{ll} 1 & |\omega| < 1 \\ |\omega|^{-n} & |\omega| > 1 \end{array} \qquad (1.2.13)$$

From this second example, direct substitution into (1.2.4) leads to difficulties in the integration and it is preferable to calculate the group delay response directly. From (1.2.4)

$$T_g(\omega_0) = \frac{-d\psi(\omega_0)}{d\omega_0} = \frac{1}{\pi} \int_{-\infty}^{\infty} \frac{\alpha(\omega)(\omega^2 + \omega_0^2)}{(\omega^2 - \omega_0^2)^2} d\omega \qquad (1.2.14)$$

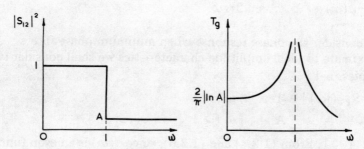

Figure 1.2.2   Group delay response of ideal amplitude filter

Substituting in the requirement (1.2.13) using (1.2.6) gives

$$T_g(\omega) = \frac{1}{\pi} \int_1^\infty n \ln y \left[\frac{1}{(y-\omega)^2} + \frac{1}{(y+\omega)^2} \right] dy \qquad (1.2.15)$$

Integrating each term by parts yields

$$T_g(\omega) = \frac{n}{\pi} \left\{ \left[\frac{-\ln y}{y+\omega} \right]_1^\infty + \left[\frac{-\ln y}{y-\omega} \right]_1^\infty + \int_1^\infty \left[\frac{1}{y(y+\omega)} + \frac{1}{y(y-\omega)} \right] dy \right\}$$

$$= \frac{n}{\omega\pi} \int_1^\infty \left(\frac{1}{y-\omega} - \frac{1}{y+\omega} \right) dy$$

or

$$T_g(\omega) = \frac{n}{\omega\pi} \ln \left| \frac{1+\omega}{1-\omega} \right| \qquad (1.2.16)$$

Although this response rises to infinity at the cut-off frequency the variation in the passband from a constant value near $\omega = 0$ is less than in the previous example. This may be seen from the power-series expansion of (1.2.16) for $|\omega| < 1$, i.e.

$$T_g(\omega) = T_{g0} \left(1 + \frac{\omega^2}{3} + \frac{\omega^4}{5} + \frac{\omega^6}{5} + \ldots \right) \qquad (1.2.17)$$

as compared to the expansion of (1.2.12)

$$T_g(\omega) = T_{g0}(1 + \omega^2 + \omega^4 + \omega^6 + \ldots) \qquad (1.2.18)$$

Thus the rate of variation of the group delay response around $\omega = 0$ increases as the rate of variation of the amplitude characteristic increases in the vicinity of the cut-off frequency.

If a non-minimum phase filter is designed with a linear phase or constant delay response in the passband and constant phase or zero delay in the stopband, the corresponding amplitude response may be calculated using (1.2.5). Thus if

$$\begin{aligned}-\psi(\omega) &= \tau\omega & |\omega|<1 \\ &= \tau \frac{\omega}{|\omega|} & |\omega|>1\end{aligned} \qquad (1.2.19)$$

from (1.2.5)

$$\alpha(\omega) = \alpha(0) - \frac{2\tau\omega^2}{\pi} \left[\int_0^1 \frac{dy}{y^2 - \omega^2} + \int_1^\infty \frac{dy}{y(y^2 - \omega^2)} \right]$$

$$= \alpha(0) + \frac{\tau}{\pi} [(1+\omega)\ln|1+\omega| + (1-\omega)\ln|1-\omega|]$$

$$\qquad (1.2.20)$$

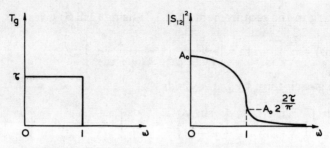

Figure 1.2.3 Amplitude response of ideal linear phase filter

and

$$\frac{d\alpha(\omega)}{d\omega} = \frac{\tau}{\pi} \ln \left| \frac{1+\omega}{1-\omega} \right| \qquad (1.2.21)$$

which deviates from the constant amplitude requirement in the passband, but the rate of change of attenuation in the vicinity of the cut-off frequency is large as shown in Figure 1.2.3.

A second example captures the behaviour for the group delay response of a finite rational lumped transfer function with constant delay in the passband, i.e.

$$T_g(\omega) = \begin{array}{ll} \tau & |\omega| < 1 \\ \tau |\omega|^{-2} & |\omega| > 1 \end{array} \qquad (1.2.22)$$

The corresponding phase response is

$$-\psi(\omega) = \begin{array}{ll} \tau\omega & |\omega| < 1 \\ 2\tau \dfrac{\omega}{|\omega|} - \dfrac{\tau}{\omega} & |\omega| > 1 \end{array} \qquad (1.2.23)$$

Substitution into (1.2.5) gives

$$\alpha(\omega) = \alpha(0) - \frac{2\tau\omega^2}{\pi} \left[\int_0^1 \frac{dy}{y^2 - \omega^2} + \int_1^\infty \frac{(2y-1)dy}{y^2(y^2 - \omega^2)} \right]$$

$$= \alpha(0) - \frac{\tau}{\pi} \left(2 + \frac{(1-\omega)^2}{\omega} \ln|1-\omega| - \frac{(1+\omega)^2}{\omega} \ln|1+\omega| \right) \qquad (1.2.24)$$

and

$$\frac{d\alpha(\omega)}{d\omega} = \frac{\tau}{\pi} \left[\frac{2}{\omega} - \frac{(1-\omega^2)}{\omega^2} \ln \left| \frac{1+\omega}{1-\omega} \right| \right] \qquad (1.2.25)$$

Expansion of (1.2.25) around $\omega = 0$ gives

$$\frac{d\alpha(\omega)}{d\omega} = \frac{\tau 4\omega}{\pi}\left(\frac{1}{3} + \frac{1}{15}\omega^2 + \frac{1}{35}\omega^4 + \ldots\right) \qquad (1.2.26)$$

and for (1.2.21)

$$\frac{d\alpha(\omega)}{d\omega} = \frac{\tau}{\pi}2\omega\left(1 + \frac{\omega^2}{3} + \frac{\omega^4}{5} + \ldots\right) \qquad (1.2.27)$$

implying a greater change in attenuation around the origin as the rate of group delay increases at the cut-off frequency.

Thus, from these examples, it is apparent that if selective linear phase characteristics are required, general non-minimum phase transfer functions must be used. However, any transfer function may be written as

$$S_{12}(p) = S'_{12}(p)\frac{H(-p)}{H(p)} \qquad (1.2.28)$$

where $S'_{12}(p)$ is a minimum phase function, $H(p)$ a Hurwitz polynomial and

$$|S_{12}(j\omega)| = |S'_{12}(j\omega)| \qquad (1.2.29)$$

$$\text{Arg } S_{12}(j\omega) = \text{Arg } S'_{12}(j\omega) - 2\tan^{-1}\left[\frac{O(\omega)}{E(\omega)}\right] \qquad (1.2.30)$$

and

$$H(j\omega) = E(\omega) + jO(\omega) \qquad (1.2.31)$$

The modification to the group delay response is

$$T_g(\omega) = T'_g(\omega) - 2\left[\frac{E(\omega)dO(\omega)/d\omega - O(\omega)dE(\omega)/d\omega}{E^2(\omega) + O^2(\omega)}\right] \qquad (1.2.32)$$

and for a finite lumped network the difference term is a finite rational function. Thus, if the amplitude response is selective, e.g. as described by (1.2.9), then this additional term may be used to equalize the group delay of the form given in (1.2.12). Around $\omega = 0$ this may be achieved with a low-degree modification term, but if constant delay is required up to band edge then, due to the singularity in the group delay response, the degree of this term will necessarily rise significantly. Thus, for a selective linear phase filter of reasonable degree, the type of characteristic which should be approximated is illustrated in Figure 1.2.4, and its realization will necessarily require a multipath non-minimum phase filter.

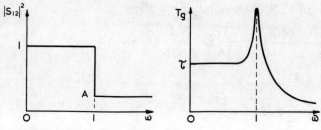

Figure 1.2.4   Response of ideal selective linear phase filter

1.3 EQUIRIPPLE RESPONSE CHARACTERISTICS

The continuous response characteristics of finite networks must approximate to the derived characteristics, possibly described in a piecewise linear manner, in some prescribed way. Consider the amplitude response of a finite lumped network of degree n and

$$F_n(\omega^2) = |S_{12}(j\omega)|^2 \tag{1.3.1}$$

is a rational function of degree n in ω^2. A normal filter specification would be as described in Figure 1.3.1 in normalized low-pass form, where $F_n(\omega^2)$ must lie within the shaded areas for $0 \leq \omega \leq 1$ and $\omega \geq \omega_1$.

$dF_n(\omega^2)/d\omega$, apart from being zero at the origin and infinity since $F_n(\omega^2)$ is a rational function of degree n in ω^2, may be zero up to $2(n-1)$ times in the interval $0 < \omega < \infty$. If these zeros occur only when $F_n(\omega^2)$ attains the limits of the specification, i.e. 1, $1 - h^2$, A and 0, then we have the equiripple solution shown for $n = 5$ in Figure 1.3.2. The same number of ripples occur in each band since the function $K - F_n(\omega^2)$ and its derivative cannot simultaneously be zero more than $n/2$, for n even, or $(n-1)/2$, for n odd, times in the interval $0 < \omega < \infty$.

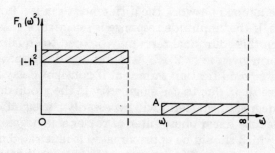

Figure 1.3.1   Method of amplitude specification

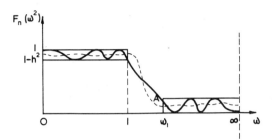

Figure 1.3.2   Equiripple amplitude response
$(n = 5)$

Consider a second solution to the approximation problem $F'_m(\omega^2)$ which meets the specification described in Figure 1.3.1. A typical solution is represented by the dashed line in Figure 1.3.2. Thus the function

$$\epsilon(\omega^2) = F_n(\omega^2) - F'_m(\omega^2) \qquad (1.3.2)$$

must be zero at least n times in the interval $0 \leqslant \omega \leqslant 1$, n times in the interval $\omega_1 \geqslant \omega \geqslant \infty$ and once in the interval $1 \geqslant \omega \geqslant \omega_1$. Thus, in total $\epsilon(\omega^2)$ must be zero at least $2n + 1$ times. Hence from (1.3.2) either $m > n$ or $\epsilon(\omega^2) \equiv 0$. Therefore, the equiripple solution is the minimum degree solution to the problem of minimizing the maximum deviation in each band of a rational function for the two band specifications described in Figure 1.3.1.

As $\omega_1 \to \infty$, $A \to 0$ and the first $2n - 1$ derivatives vanish at $\omega = \infty$ yielding a maximally flat solution around $\omega = \infty$ and

$$|S_{12}(j\omega)|^2 = \frac{1}{P_n(\omega^2)} \qquad (1.3.3)$$

where $P_n(\omega^2)$ is an nth-degree polynomial in ω^2. Using the same technique which has been applied in the general case, the equiripple solution in the band $0 \leqslant \omega \leqslant 1$ is the optimum solution to the problem of a polynomial in ω^2 constrained to lie in a band for $0 \leqslant \omega \leqslant 1$ and attain a maximum value for ω in the band $\omega_1 \leqslant \omega \leqslant \infty$ with $\omega_1 > 1$.

Dividing ω by ω_c and allowing $\omega_c \to 0$ as $h^2 \to 0$ yields a maximally flat solution at the origin with the first $2n - 1$ derivatives vanishing. This zero bandwidth approximation plays an important role as the simplest meaningful solution to the approximation problem and represents the limiting solution of most finite-band solutions, including the equiripple case, as the band of the approximation approaches zero. Consequently, it represents the basic starting point in the approximation problem for networks from which equiripple and other finite band approximations can be developed.

For a finite lumped network the phase response is of the form

$$-\psi(\omega) = \tan^{-1} \frac{O(\omega)}{E(\omega)} \tag{1.3.4}$$

where $O(\omega)/E(\omega)$ is an odd function of degree n. In an attempt to work with rational functions, the group delay function $T_g(\omega)$ is often considered, where

$$T_g(\omega) = -\frac{d\psi(\omega)}{d\omega} = \frac{E(\omega)O'(\omega) - O(\omega)E'(\omega)}{E^2(\omega) + O^2(\omega)}$$

$$= \mathrm{Ev}\left[\frac{H_n'(p)}{H_n(p)}\right]_{p=j\omega} \tag{1.3.5}$$

where the prime represents differentiation and

$$H_n(j\omega) = E(\omega) + jO(\omega) \tag{1.3.6}$$

$T_g(\omega)$ is therefore an nth-degree function in ω^2 constrained by the fact that the residue at every pole in the complex variable must be unity, since (1.3.5) may be expanded as

$$T_g(\omega) = \mathrm{Ev}\left[\sum_{r=1}^{n} \frac{1}{p - p_r}\right]_{p=j\omega} \tag{1.3.7}$$

where

$$H(p) = K \prod_{r=1}^{n} (p - p_r) \tag{1.3.8}$$

For a linear phase response in the band $0 \leq \omega \leq \omega_c$, we require a constant group delay response. Thus we could consider a specification requiring the group delay to be within the band described by the shaded area in Figure 1.3.3, noting that for lumped networks $T_g(\infty) = 0$.

Consider an equiripple solution to the problem $T_g(\omega) = F_n(\omega)$; since there are only n independent parameters we only have control over the level of $n-1$ turning points in the interval $0 \leq \omega \leq \omega_c$. Thus for $n = 6$, the equiripple solution would be as shown in Figure 1.3.4.

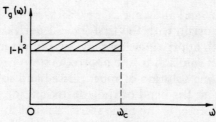

Figure 1.3.3 Method of group delay specification

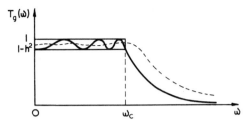

Figure 1.3.4 Equiripple group delay response

Consider a second solution to the problem where $T_g(\omega) = F_m(\omega^2)$ which is constrained within the band $1 \geqslant F'_m(\omega^2) \geqslant 1 - h^2$ for $0 \leqslant \omega \leqslant \omega_c$ and illustrated by the dashed line in Figure 1.3.4. Then

$$\epsilon(\omega^2) = F_n(\omega^2) - F'_m(\omega^2) \tag{1.3.9}$$

must be zero at least n times in the interval $0 \leqslant \omega \leqslant \omega_c$. Even though $F'_m(\omega^2)$ is constrained to have unity residues in the p variable and $F'_m(\infty) = 0$, the only necessary condition to emerge is that $m \geqslant 1$. In fact particular examples may be constructed where $m < n$ and hence the equiripple solution bears no relationship to the optimum solution of minimizing the maximum deviation of $T_g(\omega)$ in the band $0 \leqslant \omega \leqslant \omega_c$. This arises because one is trying to optimize a constrained function which may possess more turning points than the number of independent parameters which uniquely define the function. Thus, there is no significance in the equiripple group delay solution and this will not be pursued further.

Consider the phase function of the form (1.3.4) which possesses n independent parameters and is constrained to lie within the shaded region shown in Figure 1.3.5, i.e.

$$|\omega + \psi_1(\omega)| \leqslant \epsilon \qquad |\omega| \leqslant \omega_c \tag{1.3.10}$$

where

$$\psi_1(\omega) = \tan^{-1} \frac{O_1(\omega)}{E_1(\omega)} \tag{1.3.11}$$

Consider an equiripple solution to this problem as illustrated in Figure 1.3.6 for $n = 3$. Now construct a second phase function of the form

$$-\psi_2(\omega) = \tan^{-1} \frac{O_2(\omega)}{E_2(\omega)} \tag{1.3.12}$$

which is also constrained by

$$|\omega + \psi_2(\omega)| \leqslant \epsilon \qquad |\omega| \leqslant \omega_c \tag{1.3.13}$$

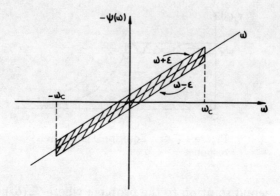

Figure 1.3.5   Method of phase specification

then

$$\epsilon(\omega) = \psi_1(\omega) - \psi_2(\omega)$$

$$= \tan^{-1} \frac{O_2(\omega)}{E_2(\omega)} - \tan^{-1} \frac{O_1(\omega)}{E_1(\omega)}$$

$$= \tan^{-1} \frac{E_1(\omega)O_2(\omega) - E_2(\omega)O_1(\omega)}{E_1(\omega)E_2(\omega) + O_1(\omega)O_2(\omega)} \qquad (1.3.14)$$

must vanish at $2n + 1$ points in the interval $-\omega_c \leq \omega \leq \omega_c$ which is impossible unless $\psi_1(\omega) = \psi_2(\omega)$ or $\tan \psi_2(\omega)$ is of degree greater than n. Thus the equiripple solution is optimum in the sense of minimizing the maximum deviation from phase linearity over a finite band. A similar result directly follows for the phase delay function

$$T_p(\omega) = -\frac{\psi(\omega)}{\omega} \qquad (1.3.15)$$

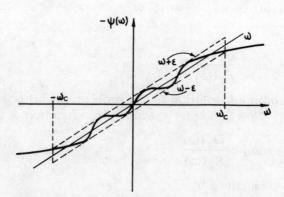

Figure 1.3.6   Equiripple linear phase response

where the equiripple solution minimizes the deviation from a constant value over a finite band for a given degree.

To solve for the equiripple linear phase polynomial $H(p)$, where

$$H(j\omega) = E(\omega) + jO(\omega) \tag{1.3.16}$$

one must make the function

$$\omega - \tan^{-1} \frac{O(\omega)}{E(\omega)} \tag{1.3.17}$$

optionally equiripple. However, there is no known analytical solution to this highly non-linear, transcendental problem (in terms of the coefficients of $H(p)$). Thus it is advantageous to seek a finite-band solution which does possess an analytical solution from which important points such as the conditions under which $H(p)$ is a Hurwitz polynomial may be extracted.

From Figure 1.3.5 it may be observed that the phase function intersects the linear characteristic at n points in the range. $0 < \omega \leqslant \omega_c$ say $\omega = \omega_i$, $i = 1 \rightarrow n$. These points of interpolation may be used as the basis of the approximation. Even for the more general problem of determining the nth-degree polynomial $H(p)$ which satisfies the constraints

$$\text{Arg } H(j\omega_i) = \psi(\omega_i) \qquad i = 1 \rightarrow n \tag{1.3.18}$$

this set of equations is a set of linear simultaneous equations in the coefficients of $H(p)$ which may readily be solved analytically. In certain cases explicit solutions may be found; one example being the equidistant interpolation to linear phase which has several important properties and is discussed in detail in Chapter 3. Even for the numerical solution to the equiripple problem, working in terms of the interpolation frequencies to the linear charateristics simplifies the process.

In the limiting case, the maximally flat solution is recovered for most of the finite-band approximations and in the linear phase case we have

$$\omega + \psi(\omega) = a_1 \omega^{2n+1} + a_2 \omega^{2n+3} + \ldots \tag{1.3.19}$$

This solution, as in the amplitude case, forms the basic one from which finite-band solutions may be developed.

CHAPTER 2

Amplitude Approximations for Lumped Networks

2.1 INTRODUCTION

For a real finite lumped network, we have

$$|S_{12}(j\omega)|^2 = \frac{\sum\limits_{1}^{m} a_i \omega^{2i}}{\sum\limits_{1}^{n} b_i \omega^{2i}} \qquad (2.1.1)$$

and our initial concern will be centred upon the derivation of a low-pass prototype response whose ideal characteristic for $\omega > 0$ is shown in Figure 2.1.1, i.e.

$$\begin{aligned}|S_{12}(j\omega)|^2 &= A & |\omega| &< \omega_c \\ |S_{12}(j\omega)|^2 &= 0 & |\omega| &> \omega_c\end{aligned} \qquad (2.1.2)$$

with $A \leqslant 1$ in order to ensure the possibility of constructing a bounded real $S_{12}(p)$. Since $|S_{12}(j\omega)|^2$ is a finite rational function in ω^2, it is continuous and therefore we may only approximate the ideal characteristic within a certain prescribed error. Thus a typical bound upon the response may be expressed as illustrated in Figure 2.1.2, where the response is constrained to lie in the shaded areas in the prescribed passband and stopband, i.e.

$$\begin{aligned}A \geqslant |S_{12}(j\omega)|^2 &\geqslant A' & |\omega| &< \omega_p \\ |S_{12}(j\omega)|^2 &\leqslant B & |\omega| &> \omega_s\end{aligned} \qquad (2.1.3)$$

We shall consider four types of solution to this approximation problem. The most general is where the response is equiripple with the maximum deviation within the prescribed limits in both the passband and stopband. This is known as the elliptic function response due to its solution being dependent upon Jacobian elliptic functions, and is treated in Section 2.5. If $\omega_p \to 0$ and $A' \to A$, then we approach a maximally flat solution about the origin. This zero bandwidth approximation is termed a maximally flat response in the passband

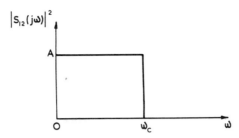

Figure 2.1.1 Ideal low-pass prototype response

since it implies that the maximum number of derivatives of $|S_{12}(j\omega)|^2$ are equated to zero at $\omega = 0$. The response with optimum equiripple response in the stopband and optimum maximally flat response in the passband is called the inverse Chebyshev response and is considered in Section 2.4. Applying the maximally flat constraint to the stopband and retaining the equiripple behaviour in the passband results in the Chebyshev response (Section 2.3). For zero bandwidth approximation in both bands i.e. a maximally flat response in both the passband and stopband we recover the maximally flat response. (Section 2.2).

Initially, in this chapter we will derive the transfer characteristic for these four cases without considering any specific realization. Then the case of the maximally flat and Chebyshev filters will be considered with regard to a formal synthesis procedure which results in explicit formulas for element values in the low-pass prototype network. Similarly, the development of a synthesis procedure for the inverse Chebyshev and elliptic function filters leads to explicit formulas for element values in the prototype network. Finally, the design of bandpass and bandstop filters is presented using a reactance slope parameter technique which is valid for a variety of physical realizations after establishing methods for determining the required degree, frequency transformation and impedance scaling.

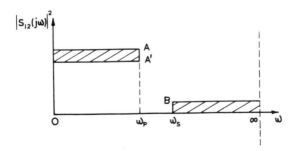

Figure 2.1.2 Lowpass prototype specification

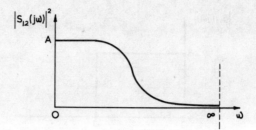

Figure 2.2.1   Maximally flat response

2.2 MAXIMALLY FLAT RESPONSE

The case of a maximally flat response is depicted in Figure 2.2.1 for an nth-degree $S_{12}(p)$.

After applying the conditions that $|S_{12}(0)|^2 = A$ and $|S_{12}(j\omega)|^2 = 0$, we equate the maximum number of derivatives to zero at both $\omega = 0$ and $\omega = \infty$. From equation (2.1.1), applying the conditions $|S_{12}(0)|^2 = A$ and $|S_{12}(j\infty)|^2 = 0$ we have

$$|S_{12}(j\omega)|^2 = A\left(\frac{1 + a_1\omega^2 + a_2\omega^4 + \ldots + a_{n-1}\omega^{2n-2}}{1 + b_1\omega^2 + b_2\omega^4 + \ldots + b_{n-1}\omega^{2n-2} + b_n\omega^{2n}}\right) \quad (b_n > 0) \quad (2.2.1)$$

Thus the maximum number of derivatives, with respect to ω, that may be equated to zero is $2n - 1$ at both $\omega = 0$ and $\omega = \infty$. To apply this condition at the origin we rewrite (2.2.1) as

$$|S_{12}(j\omega)|^2 - A = A\left[\frac{(a_1 - b_1)\omega^2 + (a_2 - b_2)\omega^4 + \ldots + (a_{n-1} - b_{n-1})\omega^{2n-2} - b_n\omega^{2n}}{1 + b_1\omega^2 + b_2\omega^4 + \ldots + b_{n-1}\omega^{2n-2} + b_n\omega^{2n}}\right] \quad (2.2.2)$$

and require that the function $|S_{12}(j\omega)|^2 - A$ be zero and its first $2n - 1$ derivatives be zero at $\omega = 0$, i.e. the power-series expansion about $\omega = 0$ should be of the form

$$|S_{12}(j\omega)|^2 - A = c_n\omega^{2n} + c_{n+1}\omega^{2n+2} + \ldots$$

which immediately implies

$$a_i = b_i \quad \text{for } i = 1 \to n - 1 \quad (2.2.3)$$

To apply the maximally flat condition at $\omega = \infty$, equation (2.2.1) is written as

$$|S_{12}(j\omega)|^2 = A\left(\frac{a_{n-1}\omega^{-2} + a_{n-2}\omega^{-4} + \ldots + a_1\omega^{-2n+2} + \omega^{-2n}}{b_n + b_{n-1}\omega^{-2} + \ldots + b_1\omega^{-2n+2} + \omega^{-2n}}\right)$$
(2.2.4)

and we require the series expansion of $|S_{12}(j\omega)|^2$ about $\omega = \infty$ to be of the form

$$|S_{12}(j\omega)|^2 = c'_n\omega^{-2n} + c'_{n+1}\omega^{-2n-2} + \ldots$$
(2.2.5)

from which we obtain

$$a_i = 0 \qquad i = 1 \to n-1$$
(2.2.6)

From (2.2.3) and (2.2.6), $a_i = b_i = 0$ for $i = 1 \to n-1$ and consequently (2.2.1) reduces to

$$|S_{12}(j\omega)|^2 = \frac{A}{1 + b_n\omega^{2n}}$$
(2.2.7)

Without changing the prescribed maximally flat behaviour, we may scale ω by any arbitrary constant and this is normally chosen such that b_n is normalized to unity resulting in

$$|S_{12}(j\omega)|^2 = \frac{A}{1 + \omega^{2n}}$$
(2.2.8)

where at $\omega = 1$ we have the half-power or 3dB point as illustrated in Figure 2.2.2.

To construct $S_{12}(p)$ from (2.2.8) we must obtain the poles of $|S_{12}(j\omega)|^2$ which occur when

$$\omega^{2n} = -1 = e^{j(2r-1)\pi} \qquad r = 1 \to 2n$$

i.e.

$$\omega = e^{j(2r-1)\pi/2n}$$

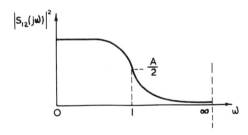

Figure 2.2.2 Normalized maximally flat response

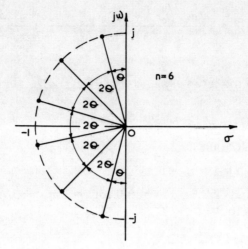

Figure 2.2.3 Pole distribution for maximally flat response

or

$$p = j\, e^{j\theta_r} \qquad r = 1 \to 2n$$
$$= -\sin\theta_r + j\cos\theta_r \qquad (2.2.9)$$

with

$$\theta_r = \frac{(2r-1)\pi}{2n}$$

To obtain a bounded real $S_{12}(p)$ the left half-plane poles are selected, i.e. for $r = 1 \to n$, to give

$$S_{12}(p) = \frac{A^{\frac{1}{2}}}{\prod_{1}^{n}(p - je^{j\theta_r})} \qquad (2.2.10)$$

where all the zeros of $s_{12}(p)$ are at infinity and the poles lie on the unit circle in the left half-plane at equal angular spacings, since for a typical pole p_r, $|p_r| = 1$ and Arg $p_r = (2r-1)\pi/2n$. This is illustrated in Figure 2.2.3.

2.3 CHEBYSHEV RESPONSE

For the equiripple passband and maximally flat stopband response we have the Chebyshev transfer characteristic depicted in Figure 2.3.1.

Taking $|S_{12}(j\omega)|^2$ as described by equation (2.2.1) we first apply the conditions that the first $2n - 1$ derivatives vanish at $\omega = \infty$, which,

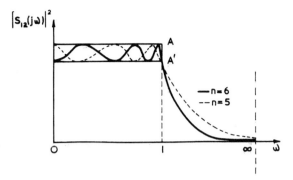

Figure 2.3.1   Chebyshev response

as in the maximally flat case, leads to

$$|S_{12}(j\omega)|^2 = \frac{C}{1 + b_1 \omega^2 + b_2 \omega^4 + \ldots + b_n \omega^{2n}} \quad (2.3.1)$$

where $C = A$ for n odd and A' for n even. Since $|S_{12}(j\omega)|^2$ never exceeds A, and also attains the value A at the maximum number of points, we may re-express $|S_{12}(j\omega)|^2$ as

$$|S_{12}(j\omega)|^2 = \frac{A}{1 + \epsilon^2 T_n^2(\omega)} \quad (2.3.2)$$

where $A' = A/(1 + \epsilon^2)$ and $T_n^2(\omega)$ is an even polynomial in ω which attains the maximum value of unity at the maximum number of points in the interval $|\omega| \leq 1$. Consequently $T_n(\omega)$ is either an even or odd polynomial in ω with the behaviour in the interval $|\omega| \leq 1$ as illustrated in Figure 2.3.2.

The problem now reduces to the determination of the polynomial $T_n(\omega)$ which is uniquely defined by its required behaviour. All points in

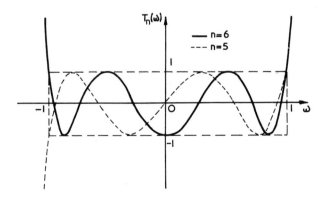

Figure 2.3.2 $T_n(\omega)$ plotted as a function of ω

the interval $|\omega|<1$ where $|T_n(\omega)|=1$ must be maximum or minimum points, i.e.

$$\frac{dT_n(\omega)}{d\omega}\bigg|_{|T_n(\omega)|=1} = 0 \qquad \text{except when } |\omega|=1 \qquad (2.3.3)$$

From this constraint we obtain the following differential equation definition of $T_n(\omega)$

$$\frac{dT_n(\omega)}{d\omega} = C_n \frac{\sqrt{1-T_n^2(\omega)}}{\sqrt{1-\omega^2}} \qquad (2.3.4)$$

Since $dT_n(\omega)/d\omega$ is a polynomial of degree $n-1$, $1-T_n^2(\omega)$ contains zeros of multiplicity 2 in $|\omega|<1$ and simple zeros at $\omega = \pm 1$, leading to the cancellation of the factor $\sqrt{1-\omega^2}$ and consequently leaving the required $(n-1)$th-degree polynomial at the right-hand side of equation (2.3.4). The only unknown at this stage is C_n, a constant, which will be determined from the condition that $T_n(\omega)$ is a polynomial of exact degree n. Writing (2.3.4) as

$$\frac{dT_n(\omega)}{\sqrt{1-T_n^2(\omega)}} = C_n \frac{d\omega}{\sqrt{1-\omega^2}} \qquad (2.3.5)$$

and integrating results in

$$\cos^{-1} T_n(\omega) = C_n \cos^{-1} \omega \qquad (2.3.6)$$

with the constant of integration being zero since $T_n(\omega)$ is an even or odd polynomial, or

$$T_n(\omega) = \cos(C_n \cos^{-1} \omega) \qquad (2.3.7)$$

We must now show that C_n may be chosen such that $T_n(\omega)$ is an nth-degree polynomial. To achieve this, it is convenient to express (2.3.7) in the parametric form,

$$T_n(\omega) = \cos C_n \theta$$
$$\omega = \cos \theta \qquad (2.3.8)$$

Since C_n is a real constant, $T_n(\omega)$ is infinity only when $\theta = j\infty$, i.e. $\omega = \infty$. Hence $T_n(\omega)$ is an entire function in ω. The zeros of $T_n(\omega)$ occur when

$$C_n \theta = \frac{(2r-1)\pi}{2} \qquad r \text{ an integer}$$

or $\qquad (2.3.9)$

$$\omega = \cos \frac{(2r-1)\pi}{2C_n} \qquad r \text{ an integer}$$

For $T_n(\omega)$ to be the required nth-degree polynomial it must be zero for n distinct values of ω, leading to the conclusion that $C_n = n$ and the zeros are given by

$$\omega_r = \cos\frac{(2r-1)\pi}{2n} \qquad r = 1 \to n \qquad (2.3.10)$$

We have demonstrated therefore, by choosing $C_n = n$, that $T_n(\omega)$ is an entire function in ω with distinct zeros at n values of ω. To complete the proof that $T_n(\omega)$ is an nth-degree polynomial, we must show that all these zeros are of multiplicity one, i.e. $T_n(\omega)$ is rational with a totality of n zeros and $T_n(\omega)|_{\omega\to\infty} \to K\omega^n$. To prove this we differentiate $T_n(\omega)$ and demonstrate that the differential is finite at the zeros, i.e.

$$\left.\frac{dT_n(\omega)}{d\omega}\right|_{\omega=\omega_r} = \left.\frac{d[\cos(n\cos^{-1}\omega)]}{d\omega}\right|_{\omega=\omega_r}$$

$$= \left.\frac{n\sin(n\cos^{-1}\omega)}{\sqrt{1-\omega^2}}\right|_{\omega=\omega_r}$$

$$= \frac{(-1)^{r+1}n}{\sin[(2r-1)\pi/2n]} \qquad (2.3.11)$$

which is finite, and by observing that $\cos(n\cos^{-1}\omega) = \cosh(n\cosh^{-1}\omega)$ the correct behaviour as $\omega \to \infty$ emerges.

Thus $T_n(\omega)$ is a polynomial of exact degree n, known as the Chebyshev polynomial of the first kind, defined by

$$T_n(\omega) = \cos(n\cos^{-1}\omega) \qquad (2.3.12)$$

with zeros at

$$\omega = \cos\frac{(2r-1)\pi}{2n} \qquad r = 1 \to n \qquad (2.3.13)$$

In this particular case, there are simpler methods of demonstrating that $T_n(\omega)$ is an nth-degree polynomial, but in the more general elliptic function case the above treatment is more appropriate and the Chebyshev case helps to highlight the various points which must be considered. One alternative is to generate the recurrence formula for $T_{n+1}(\omega)$ as

$$T_{n+1}(\omega) = \cos[(n+1)\cos^{-1}\omega]$$
$$= \omega\cos(n\cos^{-1}\omega) - \sqrt{1-\omega^2}\sin(n\cos^{-1}\omega) \qquad (2.3.14)$$

Also,

$$T_{n-1}(\omega) = \cos[(n-1)\cos^{-1}\omega]$$
$$= \omega\cos(n\cos^{-1}\omega) + \sqrt{1-\omega^2}\sin(n\cos^{-1}\omega) \qquad (2.3.15)$$

Adding (2.3.14) and (2.3.15) we have
$$T_{n+1}(\omega) = 2\omega T_n(\omega) - T_{n-1}(\omega) \tag{2.3.16}$$
and with the initial conditions
$$T_0(\omega) = 1, \qquad T_1(\omega) = \omega \tag{2.3.17}$$
$T_{n+1}(\omega)$ may be generated directly as a polynomial of degree $n + 1$.

From equation (2.3.2), we may now construct $S_{12}(p)$. All the zeros are at infinity as in the maximally flat case and the poles occur when
$$\epsilon^2 T_n^2(\omega) = -1 \tag{2.3.18}$$
Defining a new positive auxiliary parameter η as
$$\eta = \sinh\left(\frac{1}{n}\sinh^{-1}\frac{1}{\epsilon}\right) \tag{2.3.19}$$
the poles occur when
$$\cos^2(n\cos^{-1}\omega) = -\sinh^2(n\sinh^{-1}\eta)$$
$$= \sin^2(n\sin^{-1}j\eta) \tag{2.3.20}$$
Hence
$$n\cos^{-1}\omega = n\sin^{-1}j\eta + \frac{(2r-1)\pi}{2}$$
i.e.
$$p = -j\cos\left[\sin^{-1}j\eta + \frac{(2r-1)\pi}{2n}\right] \quad \text{for } r = 1 \to 2n \tag{2.3.21}$$
Choosing the left half-plane roots, the denominator of $S_{12}(p)$ is
$$K\prod_1^n\left\{p + j\cos\left[\sin^{-1}j\eta + \frac{(2r-1)\pi}{2n}\right]\right\} \tag{2.3.22}$$
A typical pole is given by
$$p_r = \sigma_r + j\omega_r = j\cos\left[\sin^{-1}j\eta + \frac{(2r-1)\pi}{2n}\right]$$
$$= \eta\sin\theta_r + j\sqrt{1+\eta^2}\cos\theta_r \tag{2.3.23}$$
with $\theta_r = (2r-1)\pi/2n$. Hence
$$\frac{\sigma_r^2}{\eta^2} + \frac{\omega_r^2}{1+\eta^2} = 1 \tag{2.3.24}$$
which is the equation of an ellipse. Multiplying ω_r by $\eta/\sqrt{1+\eta^2}$ we

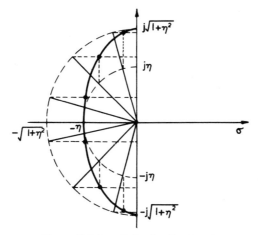

Figure 2.3.3 Pole distribution for Chebyshev response

return to the maximally flat case and therefore the pole locations for the Chebyshev case are obtained by constructing those for the maximally flat response on circles of radius $\sqrt{1+\eta^2}$ projecting parallel to the σ-axis and ω-axis respectively, and the intersection of the corresponding projections gives the required location, as shown in Figure 2.3.3.

To complete the construction of $S_{12}(p)$ we must now obtain the constant multiplier since all the zero locations (at $p = \infty$) are known. To obtain this we first note that for n even, taking complex conjugate poles, (2.3.22) may be written as

$$K \prod_{1}^{\frac{1}{2}n} (p^2 + 2\eta \sin \theta_r \, p + \eta^2 + \cos^2 \theta_r) \qquad (2.3.25)$$

and for n odd, $\frac{1}{2}n \to \frac{1}{2}(n-1)$ in (2.3.25) and an additional factor $(p + \eta)$ is obtained. For the odd case, $S_{12}(0) = A^{1/2}$, and since

$$\eta \prod_{1}^{\frac{1}{2}(n-1)} \left[\eta^2 + \cos^2 \frac{(2r-1)\pi}{2n} \right] = \eta \prod_{1}^{\frac{1}{2}(n-1)} \left[\eta^2 + \sin^2 \frac{r\pi}{n} \right]$$

$$S_{12}(p) = \frac{A^{1/2} \eta \prod_{1}^{\frac{1}{2}(n-1)} [\eta^2 + \sin^2(r\pi/n)]}{\prod_{1}^{n} \{p + j \cos [\sin^{-1} j\eta + (2r-1)\pi/2n]\}} \qquad (2.3.26)$$

For the even-degree case,

$$S_{12}(0) = \frac{A^{1/2}}{\sqrt{1+\epsilon^2}} \qquad (2.3.27)$$

which from (2.3.19) becomes

$$S_{12}(0) = A^{1/2} \tanh(n \sinh^{-1} \eta)$$

$$= \frac{A^{1/2} \prod_{1}^{\frac{1}{2}n} [\eta^2 + \sin^2(r\pi/n)]}{\prod_{1}^{\frac{1}{2}n} \{\eta^2 + \sin^2[(2r-1)/2n]\}}$$

$$= \frac{A^{1/2} \prod_{1}^{\frac{1}{2}n} [\eta^2 + \sin^2(r\pi/n)]}{\prod_{1}^{\frac{1}{2}n} \{\eta^2 + \cos^2[(2r-1)\pi/2n]\}} \quad (2.3.28)$$

Thus

$$S_{12}(p) = \frac{A^{1/2} \prod_{1}^{\frac{1}{2}n} [\eta^2 + \sin^2(r\pi/n)]}{\prod_{1}^{n} \{p + j \cos[\sin^{-1} j\eta + (2r-1)\pi/2n]\}} \quad (2.3.29)$$

leading to the general formula for n even or odd as

$$S_{12}(p) = \frac{A^{1/2} \prod_{1}^{n} [\eta^2 + \sin^2(r\pi/n)]^{1/2}}{\prod_{1}^{n} \{p + j \cos[\sin^{-1} j\eta + (2r-1)\pi/2n]\}} \quad (2.3.30)$$

2.4 INVERSE CHEBYSHEV RESPONSE

The response characteristic which is maximally flat in the passband and equiripple in the stopband, as shown in Figure 2.4.1 for $n = 6$, is the inverse Chebyshev response and may be obtained directly from the Chebyshev case as follows.

Subtract the response (2.3.2) as shown in Figure 2.3.1 from A and replace ω by $1/\omega$, i.e.

$$|S_{12}(j\omega)|^2 = A - \frac{A}{1 + \epsilon^2 T_n^2(1/\omega)}$$

$$= \frac{A}{1 + \dfrac{1}{\epsilon^2 T_n^2(1/\omega)}} \quad (2.4.1)$$

In order to use the results of the Chebyshev case directly, and also to

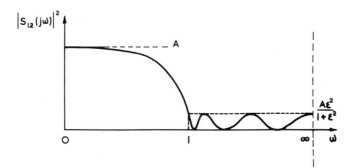

Figure 2.4.1   Inverse Chebyshev response

simplify the synthesis procedures, we shall develop the high-pass response through applying the transformation $\omega \to 1/\omega$ to give

$$|S_{12}(j\omega)|^2 = \frac{A}{1 + \frac{1}{\epsilon^2 T_n^2(\omega)}} \quad (2.4.2)$$

with the response characteristic for $n = 6$ shown in Figure 2.4.2. Writing equation (2.4.2) as

$$|S_{12}(j\omega)|^2 = \frac{A\epsilon^2 T_n^2(\omega)}{1 + \epsilon^2 T_n^2(\omega)} \quad (2.4.3)$$

we note that the denominator of $S_{12}(p)$ is identical to the Chebyshev case and the numerator is zero at the zeros of the Chebyshev polynomial. Furthermore, $S_{12}(\infty) = A^{1/2}$ and therefore

$$S_{12}(p) = \frac{A^{1/2} \prod_{1}^{n} [p + j \cos(2r - 1)\pi/2n]}{\prod_{1}^{n} \{p + j \cos[\sin^{-1} j\eta + (2r - 1)\pi/2n]\}} \quad (2.4.4)$$

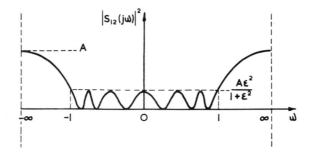

Figure 2.4.2   High-pass inverse Chebyshev response ($n = 6$)

valid for both n even and odd, giving the same pole locations as shown in Figure 2.3.3 with zeros on the imaginary axis at $\omega = \cos[(2r-1)\pi/2n]$.

2.5 ELLIPTIC FUNCTION RESPONSE

The response which is equiripple in both the passband and stopband is the elliptic function response. As was the case for the inverse Chebyshev response, it is more convenient to work with the high-pass prototype as illustrated in Figure 2.5.1 ($n = 6$).

Expressing $|S_{12}(j\omega)|^2$ in a form similar to the one for the inverse Chebyshev case we have

$$|S_{12}(j\omega)|^2 = \frac{A}{1 + \dfrac{1}{\epsilon^2 F_n^2(\omega)}} \tag{2.5.1}$$

where $F_n(\omega)$ is a rational function in ω which oscillates the maximum number at times between ± 1 for $|\omega| \leq 1$ and $|F_n(\omega)| \geq 1/m_0^{1/2}$ for $|\omega| \geq 1/m^{1/2}$ with the maximum number of ripples as shown in Figure 2.5.2 ($n = 6$).

The problem now reduces to the determination of the rational function $F_n(\omega)$ which possesses these properties, i.e. all points in $|\omega| < 1$ and $|\omega| > 1/m^{1/2}$ where $|F_n(\omega)| = 1$ and $1/m_0^{1/2}$ must be maximum or minimum points

$$\left.\frac{dF_n(\omega)}{d\omega}\right|_{|F_n(\omega)|=1} = 0 \qquad \text{except when } |\omega| = 1 \tag{2.5.2}$$

$$\left.\frac{dF_n(\omega)}{d\omega}\right|_{|F_n(\omega)|=1/m_0^{1/2}} = 0 \qquad \text{except when } |\omega| = \frac{1}{m^{1/2}}$$

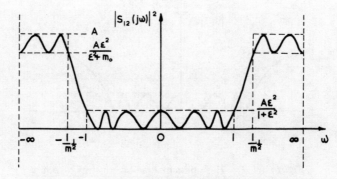

Figure 2.5.1 High-pass elliptic function response ($n = 6$)

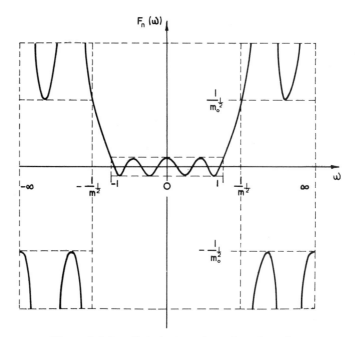

Figure 2.5.2 $F_n(\omega)$ plotted as a function of ω

From these constraints, as in the Chebyshev case, $F_n(\omega)$ is defined through the differential equation

$$\frac{dF_n(\omega)}{d\omega} = \frac{C_n\sqrt{[1-F_n^2(\omega)][1-m_0 F_n^2(\omega)]}}{\sqrt{(1-\omega^2)(1-m\omega^2)}} \quad (2.5.3)$$

Since $F_n(\omega)$ is to be a rational function of degree n, $1-F_n^2(\omega)$ and $1-m_0 F_n^2(\omega)$ will contain zeros of multiplicity 2 except at $|\omega|=1$ and $1/m^{1/2}$ leading to the correct degree on the right-hand side of the equation after the cancellation of the factors $\sqrt{1-\omega^2}$ and $\sqrt{1-m\omega^2}$. The only unknown is C_n with a possible conditional relationship between m and m_0. Writing (2.5.3) as

$$\frac{dF_n(\omega)}{\sqrt{[1-F_n^2(\omega)][1-m_0 F_n^2(\omega)]}} = \frac{C_n d\omega}{\sqrt{(1-\omega^2)(1-m\omega^2)}} \quad (2.5.4)$$

and integrating gives*

$$\mathrm{cd}_0^{-1} F_n(\omega) = C_n \mathrm{cd}^{-1}\omega = u \quad (2.5.5)$$

with the constant of integration being zero since $F_n(\omega)$ is an even or odd function and where the inverse elliptic functions (complete elliptic

*The notation for elliptic functions is the same as in Reference 2.1.

integrals) are interpreted as

$$F_n(\omega) = \mathrm{cd}_0 u$$

$$= \mathrm{cd}(u \mid m_0) \qquad (2.5.6)$$

i.e. the elliptic function cd of argument u dependent upon the elliptic parameter m_0. Similarly

$$\omega = \mathrm{cd}\left(\frac{u}{C_n}\right) = \mathrm{cd}\left(\frac{u}{C_n} \bigg| m\right) \qquad (2.5.7)$$

We must now show that the constant C_n may be chosen such that $F_n(\omega)$ is a rational function of degree n. The zeros of $F_n(\omega)$ occur when

$$\mathrm{cd}_0 u = 0$$

i.e.

$$u = (2r-1)K_0 + j2mK_0' \qquad (2.5.8)$$

where r and m are integers, and K_0 and K_0' are the quarter periods along the real and imaginary axes respectively, dependent upon the elliptic parameter m_0. From (2.5.7), the zeros in terms of ω are given by

$$\omega = \mathrm{cd}\left[\frac{(2r-1)K_0}{C_n} + \frac{j2mK_0'}{C_n}\right] \qquad (2.5.9)$$

which, for the required function, must be n in number along the real axis. Hence

$$\frac{K_0}{C_n} = \frac{K}{n}$$

and $\qquad (2.5.10)$

$$\frac{K_0'}{C_n} = K'$$

where K and K' are the quarter periods with respect to the elliptic parameter m. Thus,

$$C_n = \frac{nK_0}{K} \qquad (2.5.11)$$

and we must have the conditional requirement

$$\frac{K_0}{K_0'} = \frac{K}{nK'} \qquad (2.5.12)$$

which relates the elliptic parameters m_0 and m. Consequently, (2.5.9) reduces to

$$\omega = \text{cd}\left[\frac{(2r-1)K}{n} + j2mK'\right]$$

$$= \text{cd}\,\frac{(2r-1)K}{n} \qquad r = 1 \to n \qquad (2.5.13)$$

The poles of $F_n(\omega)$ from (2.5.6) occur when

$$u = (2r-1)K_0 + j(2m-1)K'_0$$

i.e.

$$\frac{u}{C_n} = \frac{(2r-1)K}{n} + j(2m-1)K' \qquad (2.5.14)$$

or

$$\omega = \text{cd}\left[\frac{(2r-1)K}{n} + jK'\right]$$

$$= m^{-1/2}\,\text{dc}\,\frac{(2r-1)K}{n} \qquad r = 1 \to n \qquad (2.5.15)$$

since $\text{cd}(x + jK') = m^{1/2}\,\text{dc}\,x$.

To complete the proof that $F_n(\omega)$ is a rational function of degree n, we must now show that all the zeros and poles are of multiplicity one and $F_n(\omega)$ has the correct behaviour as $\omega \to \infty$. Writing

$$F_n(\omega) = \text{cd}_0\left(\frac{nK_0}{K}\text{cd}^{-1}\,\omega\right) \qquad (2.5.16)$$

we have

$$\frac{dF_n(\omega)}{d\omega} = \frac{\text{sd}_0\left(\frac{nK_0}{K}\text{cd}^{-1}\,\omega\right)\text{nd}_0\left(\frac{nK_0}{K}\text{cd}^{-1}\,\omega\right)\frac{nK_0}{K}}{\sqrt{(1-\omega^2)(1-m\omega^2)}} \qquad (2.5.17)$$

and therefore

$$\left.\frac{dF_n(\omega)}{d\omega}\right|_{F_n(\omega)=0} = \frac{nK_0}{K(1-m)\sqrt{(1-\omega_r^2)(1-m\omega_r^2)}} \neq 0 \qquad (2.5.18)$$

since

$$|\omega_r| = \left|\text{cd}\left[\frac{(2r-1)K}{n}\right]\right| < 1$$

Similarly, differentiating $1/F_n(\omega)$ and evaluating at $1/F_n(\omega) = 0$ yields a finite value ensuring that all the poles and zeros of $F_n(\omega)$ are simple. Hence, since in addition the correct behaviour as $\omega \to \infty$ follows from (2.5.16),

$$F_n(\omega) = \frac{B \prod_{r=1}^{n} \{\omega - \text{cd}[(2r-1)K/n]\}}{\prod_{r=1}^{n} \{1 - \omega m \, \text{cd}[(2r-1)K/n]\}} \qquad (2.5.19)$$

and since $F_n(1) = 1$, the constant B is given by

$$B = \prod_{r=1}^{n} \left\{ \frac{1 - m \, \text{cd}[(2r-1)K/n]}{1 - \text{cd}[(2r-1)K/n]} \right\} \qquad (2.5.20)$$

We may now proceed to determine $S_{12}(p)$ from (2.5.1) and (2.5.16). From (2.5.19) the zeros occur when

$$\prod_{r=1}^{n} \left\{ p + j \, \text{cd}\left[\frac{(2r-1)K}{n}\right] \right\} \qquad (2.5.21)$$

is zero and the poles when

$$\epsilon^2 F_n^2(\omega) = -1$$

or

$$\text{cd}_0^2 \left(\frac{nK_0}{K} \text{cd}^{-1} \omega \right) = -\frac{1}{\epsilon^2}$$

$$= \text{sn}_0^2 \left(\frac{nK_0}{K} \text{sn}^{-1} j\eta \right) \qquad (2.5.22)$$

where the auxiliary parameter η is defined by

$$j\eta = \text{sn}\left(\frac{K}{nK_0} \text{sn}_0^{-1} \frac{j}{\epsilon}\right)$$

or, by using the Jacobi imaginary transformation, in real parametric form

$$\eta = \text{sc}(u_0 \mid 1 - m)$$
$$\frac{1}{\epsilon} = \text{sc}\left(\frac{nK_0}{K} u_0 \mid 1 - m\right) \qquad (2.5.23)$$

Solving (2.5.22)

$$\frac{nK_0}{K} \text{cd}^{-1} \omega = \frac{nK_0}{K} \text{sn}^{-1} j\eta + (2r-1)K_0 + j2mK_0'$$

or

$$p = -j\,cd\left[sn^{-1} j\eta + \frac{(2r-1)K}{n}\right] \qquad r = 1 \to n \qquad (2.5.24)$$

The location of these poles for the general case is not determined very simply from geometric considerations and will not therefore be discussed further. To complete the description of $S_{12}(p)$ given by

$$S_{12}(p) = \frac{B \prod_{r=1}^{n}\left\{p + j\,cd\left[\frac{(2r-1)K}{K}\right]\right\}}{\prod_{r=1}^{n}\left\{p + j\,cd\left[sn^{-1} j\eta + \frac{(2r-1)K}{n}\right]\right\}} \qquad (2.5.25)$$

we must determine the constant multiplier B.

For n odd,

$$S_{12}(\infty) = B = A^{1/2} \qquad (2.5.26)$$

and for n even

$$S_{12}(\infty) = B = \frac{A^{1/2}}{\sqrt{1 + m_0/\epsilon^2}}$$

$$= A^{1/2}\,nd_0\left(\frac{nK_0}{K} sn^{-1} j\eta\right)$$

$$= A^{1/2} \prod_{1}^{n}\left\{\frac{1 + \eta^2 m\,sn^2(2rK/n)}{1 + \eta^2 m\,sn^2[(2r-1)K/n]}\right\}^{1/2} \qquad (2.5.27)$$

and for n either even or odd $S_{12}(p)$ may be written as

$$S_{12}(p) = A^{1/2} \prod_{1}^{n}\left[\frac{[1 + m\eta^2\,sn^2(2rK/n)]^{1/2}\{p + j\,cd[(2r-1)K/n]\}}{\{1 + m\eta^2\,cd^2[(2r-1)K/n]\}^{1/2}\{p + j\,cd[sn^{-1} j\eta + (2r-1)K/n]\}}\right]$$

$$(2.5.28)$$

2.6 SYNTHESIS OF LADDER NETWORKS

In order to realize the resistively terminated lossless networks with transfer functions which are either maximally flat, Chebyshev, inverse Chebyshev or elliptic functions we must first form the corresponding scattering reflection coefficient $S_{11}(p)$ which we shall find, in the arbitrary gain case, is not a unique operation. For the maximally flat and Chebyshev cases, all of the zeros of $S_{12}(p)$ are at $p = \infty$ and therefore a cascade realization will result in a low-pass ladder network.

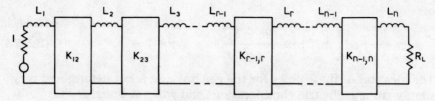

Figure 2.6.1   Modified low-pass ladder network

We shall consider these two cases first using the modified low-pass ladder network shown in Figure 2.6.1 which uses series inductors and ideal impedance inverters, the latter defined by the transfer matrix

$$\begin{bmatrix} 0 & jK \\ j/K & 0 \end{bmatrix} \tag{2.6.1}$$

with characteristic impedance K.

From equation (2.3.30), for the Chebyshev case we have

$$S_{12}(p) = A^{1/2} \prod_{1}^{n} \left\{ \frac{[\eta^2 + \sin^2(r\pi/n)]^{1/2}}{p + j\sqrt{1+\eta^2}\cos[(2r-1)\pi/2n] + \eta \sin[(2r-1)\pi/2n]} \right\} \tag{2.6.2}$$

and if we make the substitution $p \to \eta p$ and then let $\eta \to \infty$ we have

$$S_{12}(p) = \frac{A^{1/2}}{\prod_{1}^{n}[p - je^{j(2r-1)\pi/2n}]} \tag{2.6.3}$$

which is the maximally flat transfer response given in equation (2.2.10). Therefore, henceforth we shall consider only the Chebyshev case and recover the maximally flat case when required.

To form $S_{11}(p)$ we first construct $|S_{11}(j\omega)|^2$ which from equation (2.3.2) is

$$\begin{aligned} |S_{11}(j\omega)|^2 &= 1 - |S_{12}(j\omega)|^2 \\ &= \frac{1 - A + \epsilon^2 T_n^2(\omega)}{1 + \epsilon^2 T_n^2(\omega)} \end{aligned} \tag{2.6.4}$$

The denominator of $S_{11}(p)$, being necessarily devoid of zeros in Re $p \geqslant 0$, must be identical to the denominator of $S_{12}(p)$ given in equation (2.3.30). However, there is no constraint on the choice of the location of the zeros of $S_{11}(p)$ other than normally choosing complex conjugate pairs to obtain a bounded real function. We shall restrict ourselves to a minimum phase factorization, i.e. all zeros in Re $p \leq 0$, and later recover the particular case of alternating zeros in order of magnitude. For the maximum gain case ($A = 1$), all zeros are on the imaginary axis and the factorization is unique leading to a unique ladder network.

The zeros of the numerator occur when

$$T_n^2(\omega) = -\left(\frac{1-A}{\epsilon^2}\right) \qquad (2.6.5)$$

and by introducing a new auxiliary parameter*

$$\xi = \sinh\left(\frac{1}{n}\sinh^{-1}\sqrt{\frac{1-A}{\epsilon}}\right) \qquad (2.6.6)$$

we have, in a similar manner to the development of equation (2.3.22) the minimum phase factorization of $S_{11}(p)$ as,

$$S_{11}(p) = +\prod_1^n \left\{\frac{p+j\cos[\sin^{-1}j\xi + (2r-1)\pi/2n]}{p+j\cos[\sin^{-1}j\eta + (2r-1)\pi/2n]}\right\} \qquad (2.6.7)$$

the positive sign being chosen to obtain a series inductor as the first element. The input impedance

$$Z_n(p) = \frac{1+S_{11}(p)}{1-S_{11}(p)} \qquad (2.6.8)$$

is formed and will be a positive real function.

Synthesis is commenced by extracting the inductor L_1 in series to leave

$$Z'_{n-1}(p) = Z_n(p) - L_1 p \qquad (2.6.9)$$

where

$$L_1 = \left.\frac{Z_n(p)}{p}\right|_{p=\infty} > 0 \qquad (2.6.10)$$

The impedance inverter K_{12} is now extracted to leave

$$Z_{n-1}(p) = \frac{K_{12}^2}{Z'_{n-1}(p)} \qquad (2.6.11)$$

where K_{12} may be chosen as any arbitrary scaling factor which may be convenient and $Z_{n-1}(p)$ is a positive real function. The synthesis cycle may now be repeated until the remaining impedance is of zero degree and is realized as the positive load resistance R_L as shown in Figure 2.6.1.

---

*For the maximally flat case, $\xi \to \eta\alpha$ and $\eta \to \infty$ where $\alpha = (1-A)^{1/2n}$.

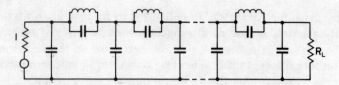

Figure 2.6.2   Network resulting from partial pole extraction technique

For the inverse Chebyshev and elliptic function responses, $S_{12}(p)$ possesses zeros on the finite real frequency axis and a conventional ladder network realization is not possible. In the odd-degree case, a partial pole extraction technique is possible with a realization as shown in Figure 2.6.2 for the low-pass response but passive elements are not always obtained. This realization has been used by Saal[2.2] and the conditions upon realizability may be obtained using Fujisawa's results.[2.3] Furthermore, to obtain the element values, numerical synthesis is required and extensive design tables.

Since we are only considering prototype network design, anticipating that one main application will be to the design of band-pass or band-stop filters, it is possible to introduce imaginary elements known as frequency invariant reactances or imaginary resistors into the prototype network. For the inverse Chebyshev and elliptic function responses this allows a ladder type of realization[2.4] as shown in Figure 2.6.3 for the high-pass prototype case. This prototype is termed the 'natural prototype' for which explicit formulas for the element values will be derived.

By allowing the elliptic parameter m to tend to zero in equation (2.5.28) we recover the inverse Chebyshev response (2.4.4) and therefore we shall consider only the elliptic function case, recovering the inverse Chebyshev response when desired.

To form $S_{11}(p)$, we have

$$|S_{11}(j\omega)|^2 = 1 - |S_{12}(j\omega)|^2$$
$$= \frac{1 + (1-A)\epsilon^2 F_n^2(\omega)}{1 + \epsilon^2 F_n^2(\omega)} \qquad (2.6.12)$$

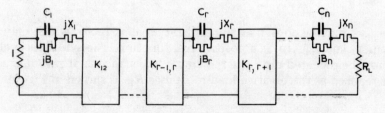

Figure 2.6.3   Natural prototype for networks with real frequency transmission zeros

from equation (2.5.1). Defining the new auxiliary parameter

$$\xi = \text{sc}(U_1 \mid 1-m) \qquad (2.6.13)$$

$$\frac{1}{\xi\sqrt{1-A}} = \text{sc}\left(\frac{nK_0}{K} U_1 \mid 1-m_0\right)$$

we may form the minimum phase factorization of $s_{11}(p)$ as

$$S_{11}(p) = B \prod_{1}^{n} \left\{ \frac{p + j\,\text{cd}[\text{sn}^{-1}\,j\xi + (2r-1)K/n]}{p + j\,\text{cd}[\text{sn}^{-1}\,j\eta + (2r-1)K/n]} \right\} \qquad (2.6.14)$$

using similar techniques to those used to derive equation (2.5.24), where B is a constant which is uniquely defined apart from a phase angle. This phase angle arises from the fact that $S_{11}(p)$ is required to be bounded but not necessarily a real function since frequency invariant reactances are to be used. At the first transmission zero,

$$\omega = -\text{cd}\,\frac{K}{n} = \omega_1 \qquad (2.6.15)$$

$|S_{11}| = 1$ since $|S_{12}| = 0$ and in particular we may choose this phase angle such that $S_{11} = 1$, resulting in

$$S_{11}(p) = \prod_{1}^{n} \left[\frac{\{p + j\,\text{cd}[\text{sn}^{-1}\,j\xi + (2r-1)K/n]\}}{\{p + j\,\text{cd}[\text{sn}^{-1}\,j\eta + (2r-1)K/n]\}} \frac{\{-j\,\text{cd}(K/n) + j\,\text{cd}[\text{sn}^{-1}\,j\eta + (2r-1)K/n]\}}{\{-j\,\text{cd}(K/n) + j\,\text{cd}[\text{sn}^{-1}\,j\xi + (2r-1)K/n]\}} \right] \qquad (2.6.16)$$

Forming the input impedance

$$Z_n(p) = \frac{1 + S_{11}(p)}{1 - S_{11}(p)} \qquad (2.6.17)$$

we have a positive function with a pole at $p = j\omega_1$. This pole is now extracted completely from $Z_n(p)$ and realized using a parallel connection of a capacitor and frequency invariant reactance in series with the input impedance to leave the remaining impedance

$$Z'_{n-1}(p) = \frac{-k_1}{p - j\omega_1} + Z_n(p) \qquad (2.6.18)$$

where k_1, the residue at $p = j\omega_1$, must be a positive real constant and $Z'_{n-1}(p)$ is a positive function.

The impedance of the parallel connection of the capacitor C_1 and frequency invariant reactance of susceptance B_1 is

$$\frac{1}{C_1 p + jB_1} \qquad (2.6.19)$$

giving

$$C_1 = \frac{1}{k_1} > 0$$

$$B_1 = \frac{\omega_1}{k_1} \quad (2.6.20)$$

Next, a frequency invariant reactance jX_1 is extracted in series from $Z'_{n-1}(p)$ such that the remaining impedance possesses a zero at the next transmission zero $p = j\omega_2$, i.e.

$$Z''_{n-1}(p) = Z'_{n-1}(p) - jX_1 \quad (2.6.21)$$

where

$$jX_1 = Z'_{n-1}(j\omega_2) \quad (2.6.22)$$

and X_1 is purely real since ω_2 is a transmission zero. Again, as in the conventional ladder case, an impedance inverter of characteristic impedance $K_{1\,2}$ is extracted to leave

$$Z_{n-1}(p) = \frac{K_{1\,2}^2}{Z''_{n-1}(p)} \quad (2.6.23)$$

where $Z_{n-1}(p)$ is a positive function and possesses a pole at $p = j\omega_2$.

The complete synthesis cycle is now repeated to generate the network shown in Figure 2.6.3, realizing all of the transmission zeros. Since the transmission zeros may be extracted in any order, the network is not unique. However, explicit formulas have only been obtained for the case where the zeros are extracted in a cyclic order defined by

$$\omega_r = -\mathrm{cd}\,\frac{(2r-1)K}{n} \quad (2.6.24)$$

and we shall concentrate on this particular configuration.

2.7 EXPLICIT FORMULAS FOR ELEMENT VALUES IN CHEBYSHEV FILTERS

The reflection coefficient for the minimum phase factorization for the Chebyshev filter is given in equation (2.6.7) and is a rational function in the three variables, p, η and ξ. Specifically recovering this three-variable dependence we have

$$S_{1\,1}(p, \eta, \xi) = \prod_{r=1}^{n} \left\{ \frac{p + j\cos[\sin^{-1}j\xi + (2r-1)\pi/2n]}{p + j\cos[\sin^{-1}j\eta + (2r-1)\pi/2n]} \right\} \quad (2.7.1)$$

and from (2.3.30)

$$S_{12}(p, \eta, \xi) = A^{1/2} \prod_{r=1}^{n} \left\{ \frac{[\eta^2 + \sin^2(r\pi/n)]^{1/2}}{p + j\cos[\sin^{-1}j\eta + (2r-1)\pi/2n]} \right\} \quad (2.7.2)$$

Now

$$A(\eta, \xi) = 1 - \epsilon^2 \sinh^2(n \sinh^{-1} \xi)$$

$$= \frac{\sinh^2(n \sinh^{-1} \eta) - \sinh^2(n \sinh^{-1} \xi)}{\sinh^2(n \sinh^{-1} \eta)} \quad (2.7.3)$$

by using equation (2.3.19). The zeros of the numerator in terms of η and ξ occur when

$$n \sinh^{-1} \eta = n \sinh^{-1} \xi + \frac{jr\pi}{n} \qquad r = 1 \to 2n \quad (2.7.4)$$

or

$$\eta = \sinh\left[\sinh^{-1}\xi + j\left(\frac{r\pi}{n}\right)\right]$$

and therefore

$$A(\eta, \xi) = \prod_{r=1}^{2n} \left[\frac{\eta - \sinh(\sinh^{-1}\xi + j\, r\pi/n)}{\eta - \sinh(j\, r\pi/n)} \right]$$

$$= \prod_{r=1}^{2n} \left[\frac{\eta - j\sin(-\sin^{-1}j\xi + r\pi/n)}{\eta - j\sin(r\pi/n)} \right] \quad (2.7.5)$$

and by collecting complex conjugate factors e.g.

$$\left[\eta - j\sin\left(-\sin^{-1}j\xi + \frac{r\pi}{n}\right)\right]\left[\eta + j\sin\left(+\sin^{-1}j\xi + \frac{r\pi}{n}\right)\right]$$

$$= \left(\eta - \xi\cos\frac{r\pi}{n} - j\sqrt{1+\xi^2}\sin\frac{r\pi}{n}\right)$$

$$\left(\eta - \xi\cos\frac{r\pi}{n} + j\sqrt{1+\xi^2}\sin\frac{r\pi}{n}\right)$$

$$= \eta^2 + \xi^2 - 2\eta\xi\cos\frac{r\pi}{n} + \sin^2\frac{r\pi}{n} \quad (2.7.6)$$

we have

$$A(\eta,\xi) = \prod_{r=1}^{n}\left[\frac{\eta^2 + \xi^2 - 2\eta\xi\cos(r\pi/n) + \sin^2(r\pi/n)}{\eta^2 + \sin^2(r\pi/n)}\right] \quad (2.7.7)$$

Substitution into (2.7.2) gives

$$S_{12}(p,\eta,\xi) = \prod_{r=1}^{n} \left\{ \frac{[\eta^2 + \xi^2 - 2\eta\xi \cos(r\pi/n) + \sin^2(r\pi/n)]^{1/2}}{p + j \cos[\sin^{-1} j\eta + (2r-1)\pi/2n]} \right\}$$

(2.7.8)

Before attempting to determine explicit formulas arising in the synthesis procedure it is necessary to obtain certain properties of these generalized three-variable rational Chebyshev functions. The two most important are obtained in the following manner.

Property 1
Defining

$$j\eta_u = \sin\left(\sin^{-1} j\eta + \frac{u\pi}{n}\right) \qquad (2.7.9)$$

then

$$S_{11}(p,\eta,\eta_u) = \prod_{r=1}^{u} \left\{ \frac{p - j \cos[\sin^{-1} j\eta + (2r-1)\pi/2n]}{p + j \cos[\sin^{-1} j\eta + (2r-1)\pi/2n]} \right\} \qquad (2.7.10)$$

which is a function of degree u in p. Similarly, $S_{11}(p,\eta,\eta_{-u})$ is a function of degree u in p.

Proof
Replacing ξ by η_u in equation (2.7.1) we have

$$S_{11}(p,\eta,\eta_u) = \prod_{r=1}^{n} \left\{ \frac{p + j \cos[\sin^{-1} j\eta + (2r + 2u - 1)/2n]}{p + j \cos[\sin^{-1} j\eta + (2r-1)\pi/2n]} \right\}$$

$$= \frac{\prod_{r=1}^{n-u} \{p + j \cos[\sin^{-1} j\eta + (2r + 2u - 1)\pi/2n]\} \prod_{1}^{u} \{p - j \cos[\sin^{-1} j\eta + (2r-1)\pi/2n]\}}{\prod_{r=1}^{u} \{p + j \cos[\sin^{-1} j\eta + (2r-1)\pi/2n]\} \prod_{1}^{n-u} \{p + j \cos[\sin^{-1} j\eta + (2r + 2u - 1)\pi/2n]\}}$$

$$= \prod_{r=1}^{u} \left\{ \frac{p - j \cos[\sin^{-1} j\eta + (2r-1)\pi/2n]}{p + j \cos[\sin^{-1} j\eta + (2r-1)\pi/2n]} \right\} \qquad (2.7.11)$$

Similarly, for u negative (2.7.11) is inverted.

Property 2
Using equation (2.7.8)

$$S_{12}\left(-j\cos\frac{\pi}{2n},\eta,0\right) = \prod_{r=1}^{n-1}\left[\frac{\sin(r\pi/n)-j\eta}{\sin(r\pi/n)+j\eta}\right]^{1/2} \tag{2.7.12}$$

Proof
From equation (2.3.2) for $A = 1$ ($\xi = 0$) and evaluating at the first zero of $T_n(\omega)$, i.e. $\omega = -\cos(\pi/2n)$, $|S_{12}|^2 = 1$ for η real. Under these conditions, S_{12} is infinite when

$$\cos\frac{\pi}{2n} = \cos\left[\sin^{-1}j\eta + \frac{(2r-1)\pi}{2n}\right] \tag{2.7.13}$$

i.e.

$$\eta = j\sin\frac{(r-1)\pi}{n} \tag{2.7.14}$$

Using (2.7.8) we have

$$S_{12}\left(-j\cos\frac{\pi}{2n},\eta,0\right) = \prod_{r=1}^{n}\left\{\frac{[\eta^2 + \sin^2(r\pi/n)]^{1/2}}{\eta - j\sin(r\pi/n)}\right\}$$

$$= \prod_{r=1}^{n}\left[\frac{\sin(r\pi/n)-j\eta}{\sin(r\pi/n)+j\eta}\right]^{1/2} \tag{2.7.15}$$

We are now in a position to determine the explicit formulas for the element values in the network.

Consider $S_{11}(p,\eta,\xi)$ defined in equation (2.7.1) and construct the input impedance of the network normalized to a generator of unity impedance, i.e.

$$Z_n(p,\eta,\xi) = \frac{1 + S_{11}(p,\eta,\xi)}{1 - S_{11}(p,\eta,\xi)} \tag{2.7.16}$$

and is a rational positive real function in p for η and ξ real under the necessary condition $\eta > \xi > 0$, from equations (2.3.19) and (2.6.6), and (2.6.6), and of degree n in p, η and ξ. Furthermore, from (2.7.1)

$$Z_n(p,\eta,\xi) = -Z_n(p,\xi,\eta)$$
$$Z_n(p,\eta,\eta) = \infty \tag{2.7.17}$$
$$Z_n(\infty,\eta,\xi) = \infty$$

Thus, completely extracting the pole at $p = \infty$ we have

$$Z'_{n-1}(p,\eta,\xi) = Z_n(p,\eta,\xi) - A_1(\eta,\xi)p \tag{2.7.18}$$

where $A_1(\eta,\xi) = L_1$ is a real positive rational function in η and ξ for

$\eta > \xi > 0$. Additionally, from (2.7.17)

$$A_1(\eta,\xi) = -A_1(\xi,\eta)$$
$$A_1(\eta,\eta) = \infty \qquad (2.7.19)$$

and

$$Z'_{n-1}(p,\eta,\xi) = -Z_{n-1}(p,\xi,\eta)$$
$$Z'_{n-1}(p,\eta,\eta) = \infty \qquad (2.7.20)$$

where $Z'_{n-1}(p,\eta,\xi)$ is a positive real function. Since all of the transmission zeros are at $p = \infty$, we also must have

$$Z'_{n-1}(\infty,\eta,\epsilon) = 0 \qquad (2.7.21)$$

An impedance inverter of characteristic impedance $K_{1\,2}(\eta,\xi)$ is now extracted to give

$$Z_{n-1}(p,\eta,\xi) = \frac{K_{1\,2}^2(\eta,\xi)}{Z'_{n-1}(p,\eta,\xi)} \qquad (2.7.22)$$

At the zeros of $K_{1\,2}(\eta,\xi)$, $Z_{n-1}(p,\eta,\xi)$ vanishes and consequently $Z_n(p,\eta,\xi)$ will be of degree one in p. From property 1, equations (2.7.9) and (2.7.10), the zeros of $K_{1\,2}^2(\eta,\xi)$ are therefore the zeros of

$$\left[\xi + j\sin\left(\sin^{-1} jn + \frac{\pi}{n}\right)\right]\left[\xi + j\sin\left(\sin^{-1} jn - \frac{\pi}{n}\right)\right]$$
$$= \xi^2 + \eta^2 - 2\eta\xi\cos\frac{\pi}{n} + \sin^2\frac{\pi}{n} \qquad (2.7.23)$$

and by deliberately enforcing the condition $K_{1\,2}(\eta,\eta) = \infty$ we have

$$K_{1\,2}(\eta,\xi) = \frac{\sqrt{\xi^2 + \eta^2 - 2\eta\xi\cos(\pi/n) + \sin^2(r\pi/n)}}{\eta - \xi} \qquad (2.7.24)$$

Thus, $Z_{n-1}(p,\eta,\xi)$ is of degree $n-1$ in p, rational in η and ξ, and

$$Z_{n-1}(p,\eta,\xi) = -Z_{n-1}(p,\xi,\eta)$$
$$Z_{n-1}(p,\eta,\eta) = \infty$$
$$Z_{n-1}(\infty,\eta,\xi) = \infty \qquad (2.7.25)$$

and $Z_{n-1}(p,\eta,\xi)$ is a positive real function if $\eta > \xi > 0$.

We may now proceed to the rth cycle of the synthesis procedure, and

$$Z_{n+1-r}(p,\eta,\xi) = A_r(\eta,\xi)p + \frac{K_{r,r+1}^2(\eta,\xi)}{Z_{n-1}(p,\eta,\xi)} \qquad (2.7.26)$$

where $A_r(\eta,\epsilon) = L_r$ is a rational function and

$$A_r(\eta,\xi) = -A_r(\xi,\eta)$$
$$A_r(\eta,\eta) = \infty \qquad (2.7.27)$$

the zeros of $K_{r,r+1}^2(\eta,\xi)$ being deliberately chosen such that $Z_n(p,\eta,\xi)$ would be of degree r in p at these points. Thus from property 1, these zeros are the zeros of

$$\left[\xi + j\sin\left(\sin^{-1} j\eta + \frac{r\pi}{n}\right)\right]\left[\xi + j\sin\left(\sin^{-1} j\eta - \frac{r\pi}{n}\right)\right]$$

$$= \xi^2 + \eta^2 - 2\eta\xi \cos\frac{r\pi}{n} + \sin^2\frac{r\pi}{n} \qquad (2.7.28)$$

and enforcing the condition $K_{r,r+1}(\eta,\eta) = \infty$, we have

$$K_{r,r+1}(\eta,\xi) = \frac{\sqrt{\xi^2 + \eta^2 - 2\eta\xi \cos(r\pi/n) + \sin^2(r\pi/n)}}{\eta - \xi} \qquad (2.7.29)$$

Furthermore, it readily follows that

$$Z_{n-r}(p,\eta,\xi) = -Z_{n-r}(p,\xi,\eta)$$
$$Z_{n-r}(p,\eta,\eta) = \infty \qquad (2.7.30)$$
$$Z_{n-r}(\infty,\eta,\xi) = \infty$$

with $Z_{n-r}(p,\eta,\xi)$ a positive real function for $\eta > \xi > 0$.

After n cycles of this procedure we are left with the terminating resistor $R_L(\eta,\epsilon)$, where

$$R_L(\eta,\eta) = \infty \qquad (2.7.31)$$

Additionally, from (2.7.1), $|S_{11}(j\omega,\eta,-\eta)| = 1$, and since the realization is in the form of a resistively terminated passive lossless two-port

$$R_L(\eta,-\eta) = 0 \quad \text{or} \quad \infty \qquad (2.7.32)$$

Up to this point no mention has been made of the degree of $A_r(\eta,\xi)$ in η and ξ. However, this is not necessary at this stage for, from the results which will be obtained in the 'matched' case ($A = 1$ or $\xi = 0$), this fact may readily be recovered. Concentrating on the matched case $\xi = 0$, we now use the property 2 given in equation (2.7.12) to obtain the coefficients $A_r(\eta,0)$ and $R_L(\eta,0)$ in a very direct and simple manner.

Using property 2 we have

$$S_{12}\left(-j\cos\frac{\pi}{2n}, -j\lambda, 0\right) = \prod_1^n \left[\frac{\sin(r\pi/n) - \lambda}{\sin(r\pi/n) + \lambda}\right]^{1/2} \qquad (2.7.33)$$

which is immediately recognizable as a bounded real all-pass function in λ. Rewriting (2.7.33) as

$$S_{12}(-j\cos\frac{\pi}{2n}, -j\lambda, 0) = \prod_1^n \frac{[\sin^2(r\pi/n) - \lambda^2]^{1/2}}{\sin(r\pi/n) + \lambda} \qquad (2.7.34)$$

we may deduce that each of these all-pass factors may be associated with a linear degree section in λ. Furthermore from broad-band matching theory of passive networks,[2.5] every secton in the two-port cascade must be passive and all-pass if the overall network is passive and all-pass. Thus, the only non-unique factor in the realization of this all-pass function is the ordering of the transmission zero producing sections along the network. Since the rth impedance inverter produces a half-ordered transmission zero at

$$\lambda = \pm \sin \frac{r\pi}{n} \qquad (2.7.35)$$

from (2.7.29), the ordering of the zeros is unique. Thus, for $p = -j\cos(\pi/2n)$, $\xi = 0$, the overall two-port network must possess a transfer matrix which decomposes into the product

$$\prod_{1}^{n-1} \frac{1}{\sqrt{\eta^2 + \sin^2(r\pi/n)}} \begin{bmatrix} \sin(r\pi/n) & j\eta \\ j\eta & \sin(r\pi/n) \end{bmatrix} \qquad (2.7.36)$$

and also

$$R_L(\eta,\infty) = 1 \qquad (2.7.37)$$

Each basic section of (2.7.36) may now be decomposed as

$$\begin{bmatrix} 1 & \dfrac{-j\sin(r\pi/n)}{\eta} \\ 0 & 1 \end{bmatrix} \begin{bmatrix} 0 & \dfrac{j\sqrt{\eta^2 + \sin^2(r\pi/n)}}{\eta} \\ \dfrac{j\eta}{\sqrt{\eta^2 + \sin^2(r\pi/n)}} & 0 \end{bmatrix}$$

$$\begin{bmatrix} 1 & \dfrac{-j\sin(r\pi/n)}{\eta} \\ 0 & 1 \end{bmatrix} \qquad (2.7.38)$$

with the centre matrix being the transfer matrix of the impedance inverter $K_{r,\,r+1}$. Thus, between the impedance inverters of characteristic impedances $K_{r-1,\,r}$ and $K_{r,\,r+1}$ we have a series element of impedance

$$\frac{-j}{\eta}\left[\sin\frac{(r-1)\pi}{n} + \sin\frac{r\pi}{n}\right] \qquad (2.7.39)$$

which is equal to $A_r(\eta,0)p$ evaluated at $p = -j\cos(\pi/2n)$. Hence

$$A_r(\eta,0) = \frac{\sin[(r-1)\pi/n] + \sin(r\pi/n)}{\eta\cos(\pi/2n)} = \frac{2}{\eta}\sin\frac{(2r-1)\pi}{2n}.$$

$$(2.7.40)$$

We may now proceed to the arbitrary gain case. From (2.7.8) it is apparent that the load resistance $R_L(\eta,\xi)$ can only be zero or infinite when $\xi = \eta$ as defined in equation (2.7.9). However, from the limiting value $R_L(\eta,0) = 1$, the only possible factors are $(\eta + \xi)$ and $(\eta - \xi)$. Thus from (2.7.31) and (2.7.32)

$$R_L(\eta,\xi) = \frac{\eta + \xi}{\eta - \xi} \qquad (2.7.41)$$

Furthermore, if $A_r(\eta,\xi)$ is zero then $S_{12}(p,\eta,\xi)$ as given in equation (2.7.8), must be of lower degree in p than n. This can only occur if $\eta = \infty$ or $\xi = \infty$ or $\xi = \eta_u$. Thus, from the limiting value (2.7.40) and condition (2.7.27) it follows that

$$A_r(\eta,\xi) = \frac{2 \sin[(2r-1)\pi/2n]}{\eta - \xi} \qquad (2.7.42)$$

We have thus established the explicit formulas for element values in the low-pass prototype networks which exhibit either a Chebyshev or maximally flat response for the case of minimum phase factorization of S_{11}. Where the zeros of S_{11} are chosen to alternate, i.e. the zeros of

$$\sqrt{1-A} + j\epsilon T_n(jp) \qquad (2.7.43)$$

the results are given in the following section but the proof is left to the interested reader since this case is not of great practical significance.

The historical development of these explicit formulas is interesting. The maximally flat formulas were first discovered in the 1930s,[2.6, 2.7] and the Chebyshev ones in the 1950s.[2.8–2.11] However, a complete proof was first obtained by Takahasi[2.12, 2.13] and differs considerably from the approach adopted in this section.

2.8 SUMMARY OF RESULTS FOR CHEBYSHEV AND MAXIMALLY FLAT FILTERS

With reference to Figure 2.8.1 where $K_{r,r+1}$ is the characteristic impedance of the inverter located between the series inductors L_r and

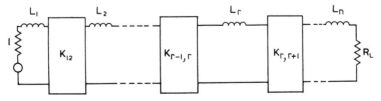

Figure 2.8.1 Prototype for maximally flat and Chebyshev filters

L_{r+1} we have:

A. The Chebyshev Prototype

1. Minimum Phase $S_{11}(p)$

$$L_r = \frac{2\sin[(2r-1)\pi/2n]}{\eta - \xi} \qquad r = 1 \to n$$

$$K_{r,\,r+1} = \frac{\sqrt{\xi^2 + \eta^2 - 2\eta\xi\cos(r\pi/n) + \sin^2(r\pi/n)}}{\eta - \xi} \qquad r = 1 \to n-1$$

(2.8.1)

$$R_L = \frac{\eta + \xi}{\eta - \xi}$$

where the auxiliary parameters η and ξ are given by

$$\eta = \sinh\left(\frac{1}{n}\sinh^{-1}\frac{1}{\epsilon}\right)$$

$$\xi = \sinh\left(\frac{1}{n}\sinh^{-1}\frac{\sqrt{1-A}}{\epsilon}\right) \qquad (2.8.2)$$

2. Matched Case ($A = 1$, $\xi = 0$, $R_L = 1$)

$$L_r = \frac{2\sin[(2r-1)\pi/2n]}{\eta} \qquad r = 1 \to n \qquad (2.8.3)$$

$$K_{r,\,r+1} = \frac{\sqrt{\eta^2 + \sin^2(r\pi/n)}}{\eta} \qquad r = 1 \to n-1$$

3. Zeros of $S_{11}(p)$ alternating between the two half-planes

The element values are the same as in the matched case with the following exceptions.

For n odd a transformer is required in the centre of the network bisecting the central inductor of turns ratio

$$n = \frac{1 - \sqrt{1-A}}{A^{1/2}} \qquad (2.8.4)$$

and for n even, the same transformer is required to bisect the central impedance inverter into two 45° phase shifters which possess transfer matrices of the form

$$\frac{1}{\sqrt{2}} \begin{bmatrix} 1 & \dfrac{j\sqrt{1+\eta^2}}{\eta} \\ \dfrac{j\eta}{\sqrt{1+\eta^2}} & 1 \end{bmatrix} \qquad (2.8.5)$$

4. *Zero Impedance Generator*

For the singly terminated case, the network is scaled by $(\eta - \xi)/(\eta + \xi)$ and ξ is allowed to approach η converting $S_{12}(p)$ into $Z_{12}(p)$ the singly terminated transfer impedance ($R_L = 1$).

$$L_r = \frac{\sin[(2r-1)\pi/2n]}{\eta} \qquad r = 1 \to n \qquad (2.8.6)$$

$$K_{r,\,r+1} = \frac{\sin(r\pi/2n)\sqrt{\eta^2 + \cos^2(r\pi/2n)}}{\eta} \qquad r = 1 \to n-1$$

B. **The Maximally Flat Prototype** ($p \to \eta p,\ \xi \to \alpha\eta,\ \eta \to \infty$)

1. *Minimum Phase* $S_{11}(p)$

$$L_r = \frac{2\sin[(2r-1)\pi/2n]}{1-\alpha} \qquad r = 1 \to n \qquad (2.8.7)$$

$$K_{r,\,r+1} = \frac{\sqrt{1 - 2\alpha \cos(r\pi/n) + \alpha^2}}{1-\alpha} \qquad r = 1 \to n-1$$

$$R_L = \frac{1+\alpha}{1-\alpha} \qquad (2.8.8)$$

$$\alpha = (1-A)^{1/2n}$$

2. *Matched Case* ($A = 1,\ \alpha = 0,\ R_L = 1$)

$$L_r = 2\sin\frac{(2r-1)\pi}{2n} \qquad r = 1 \to n \qquad (2.8.9)$$

$$K_{r,\,r+1} = 1 \qquad r = 1 \to n-1$$

3. Zeros of $S_{11}(p)$ alternating

The element values are the same as the matched case with a transformer of turns ratio

$$n = \frac{1 - \sqrt{1-A}}{A^{1/2}}$$

bisecting the central element.

4. Zero Impedance Generator ($R_L = 1, S_{12} \to Z_{12}$)

$$L_r = \sin \frac{(2r-1)\pi}{2n} \quad r = 1 \to n$$

$$K_{r, r+1} = \sin \frac{r\pi}{2n} \quad r = 1 \to n-1$$

(2.8.10)

2.9 EXPLICIT FORMULAS FOR ELEMENT VALUES IN ELLIPTIC FUNCTION FILTERS[2.4]

We shall again consider the case where $S_{11}(p)$ is a minimum phase function and closely follow the procedure for the Chebyshev case. From equation (2.6.16) we have $S_{11}(p,\eta,\xi)$ specifically written as a function of these variables as,

$$S_{11}(p,\eta,\xi) = \prod_1^n \left[\frac{\{p + j\,cd[sn^{-1}\,j\xi + (2r-1)K/n]\}\{-j\,cd(K/n) + j\,cd[sn^{-1}\,j\eta + (2r-1)K/n)]\}}{\{p + j\,cd[sn^{-1}\,j\eta + (2r-1)K/K]\}\{-j\,cd(K/n) + j\,cd[sn^{-1}\,j\xi + (2r-1)K/n]\}} \right]$$

(2.9.1)

and from (2.5.28)

$$S_{12}(p,\eta,\xi) = A^{1/2} \prod_1^n \frac{[1 + m\eta^2\,sn^2(2rK/n)]^{1/2}\{p + j\,cd[(2r-1)K/n]\}}{\{1 + m\eta^2\,cd^2[(2r-1)K/n]\}\{p + j\,cd[sn^{-1}\,j\eta + (2r-1)K/n]\}}$$

(2.9.2)

where

$$A = (\eta,\xi) = 1 + \frac{1}{\epsilon^2\,sn_0^2[n(K_0/K)sn^{-1}\,j\xi]}$$

$$= 1 - \frac{sn_0^2[n(K_0/K)sn^{-1}\,j\eta]}{sn_0^2[n(K_0/K)sn^{-1}\,j\xi]}$$

(2.9.3)

using equations (2.5.23) and (2.6.13). The zeros of the numerator occur when

$$\frac{nK_0}{K} \operatorname{sn}^{-1} j\eta = \frac{nK_0}{K} \operatorname{sn}^{-1} j\xi + j2rK_0$$

or

$$j\eta = \operatorname{sn}\left(\operatorname{sn}^{-1} j\xi + j\frac{2rK}{n}\right) \quad (2.9.4)$$

and of the denominator when

$$\operatorname{sn}_0^2 \left(\frac{nK_0}{K} \operatorname{sn}^{-1} j\eta\right) = \infty$$

and (2.9.5)

$$\operatorname{sn}_0^2 \left(\frac{nK_0}{K} \operatorname{sn}^{-1} j\xi\right) = 0$$

i.e.

$$j\eta = \operatorname{sn}\left(\frac{2rK}{n} + jK'\right)$$

$$= \frac{\operatorname{ns}(2rK/n)}{m^{1/2}}$$

and

$$j\xi = \operatorname{sn}\frac{2rK}{n} \quad (2.9.6)$$

Hence

$$A(\eta,\xi) = \prod_{1}^{n} \left\{ \frac{\xi^2[1 + m\eta^2 \operatorname{sn}^2(2rK/n)] - 2\eta\xi \operatorname{cn}(2rK/n)\operatorname{dn}(2rK/n) + \eta^2 + \operatorname{sn}^2(2rK/n)}{[\xi^2 + \operatorname{sn}^2(2rK/n)][1 + m\eta^2 \operatorname{sn}^2(2rK/n)]} \right\}$$

(2.9.7)

since $A(\eta,\infty) = 1$, and

$$S_{12}(p,\eta,\xi) = \prod_{1}^{n} \left[\frac{\{\xi^2[1 + m\eta^2 \operatorname{sn}^2(2rK/n)] - 2\eta\xi \operatorname{cn}(2rK/n)\operatorname{dn}(2rK/n) + \eta^2 + \operatorname{sn}^2(2rK/n)\}^{1/2}\{p + j \operatorname{cd}[(2r-1)K/n]\}}{[\xi^2 + \operatorname{sn}^2(2rK/n)]^{1/2}\{1 + m\eta^2 \operatorname{cd}^2[(2r-1)K/n]\}^{1/2} \{p + j \operatorname{cd}[\operatorname{sn}^{-1} j\eta + (2r-1)K/n]\}} \right]$$

(2.9.8)

We shall now develop some important properties of these generalized three-variable rational elliptic functions.

Property 1
Defining η_u as

$$j\eta_u = \mathrm{sn}\left[\mathrm{sn}^{-1} j\eta + \frac{2uK}{n}\right] \qquad (2.9.9)$$

we have

$$S_{11}(p,\eta,\eta_u) = \prod_{r=1}^{u} \left[\frac{\{p - j\,\mathrm{cd}[\mathrm{sn}^{-1} j\eta + (2r-1)K/n]\}}{\{p + j\,\mathrm{cd}[\mathrm{sn}^{-1} j\eta + (2r-1)K/n]\}} \frac{\{-j\,\mathrm{cd}(K/n) + j\,\mathrm{cd}[\mathrm{sn}^{-1} j\eta + (2r-1)K/n]\}}{\{-j\,\mathrm{cd}(k/n) - j\,\mathrm{cd}[\mathrm{sn}^{-1} j\eta + (2r-1)K/n]\}}\right]$$

$$(2.9.10)$$

which is a function of degree u in p. Also $S_{11}(p,\eta,\eta_{-u})$ is a function of degree u in p.

Proof
Consider the function

$$Q_n(p,\eta,\xi) = \prod_{r=1}^{n} \left\{\frac{p + j\,\mathrm{cd}[\mathrm{sn}^{-1} j\xi + (2r-1)K/n]}{p + j\,\mathrm{cd}[\mathrm{sn}^{-1} j\eta + (2r-1)K/n]}\right\} \qquad (2.9.11)$$

evaluated at $\xi = \eta_u$ as given in equation (2.9.9) to give

$$Q_n(p,\eta,\eta_u) = \prod_{1}^{n} \left\{\frac{p + j\,\mathrm{cd}[\mathrm{sn}^{-1} j\eta + (2r + 2u - 1)K/n]}{p + j\,\mathrm{cd}[\mathrm{sn}^{-1} j\eta + (2r - 1)K/n]}\right\}$$

$$= \frac{\displaystyle\prod_{r=1}^{n-u}\{p + j\,\mathrm{cd}[\mathrm{sn}^{-1} j\eta + (2r + 2u - 1)K/n]\} \displaystyle\prod_{r=1}^{u}\{p - j\,\mathrm{cd}[\mathrm{sn}^{-1} j\eta + (2r - 1)K/n]\}}{\displaystyle\prod_{r=1}^{u}\{p + j\,\mathrm{cd}[\mathrm{sn}^{-1} j\eta + (2r - 1)K/n]\} \displaystyle\prod_{r=1}^{n-u}\{p + j\,\mathrm{cd}[\mathrm{sn}^{-1} j\eta + (2r + 2u - 1)K/n]\}}$$

$$= \prod_{r=1}^{u}\left\{\frac{p - j\,\mathrm{cd}[\mathrm{sn}^{-1} j\eta + (2r-1)K/n]}{p + j\,\mathrm{cd}[\mathrm{sn}^{-1} j\eta + (2r-1)K/n]}\right\} \qquad (2.9.12)$$

from which (2.9.10) follows immediately since

$$S_{11}(p,\eta,\xi) = \frac{Q_n(p,\eta,\xi)}{Q_n(-j\,\mathrm{cd}[K/n],\eta,\xi)}$$

Similarly, for $\xi = \eta_{-u}$, (2.9.10) is inverted.

Property 2

$$S_{11}\left(-j\,cd\,\frac{(2q+1)}{n},\eta,\xi\right)$$

$$=\prod_{1}^{q}\left\{\frac{[\eta-j\,sn(2rK/n)][\xi+j\,sn(2rK/n)]}{[\eta+j\,sn(2rK/n)][\xi-j\,sn(2rK/n)]}\right\} \qquad (2.9.13)$$

which is a rational function of degree q in η and ξ.

Proof
Consider the function

$$V_n(p,\eta)=\prod_{r=1}^{n}\left\{\frac{-j\,cd(K/n)+j\,cd[sn^{-1}\,j\eta+(2r-1)K/n]}{p+j\,cd[sn^{-1}\,j\eta+(2r-1)K/n]}\right\} \qquad (2.9.14)$$

and evaluate at $p=-j\,cd[(2q+1)K/n]$.
The zeros of the denominator in η occur when

$$sn^{-1}\,j\eta+\frac{(2r-1)K}{n}=\frac{(2q+1)K}{n}$$

i.e.

$$\eta=j\,sn\,\frac{2(r-q-1)K}{n} \qquad (2.9.15)$$

and those of the numerator when

$$\eta=j\,sn\,\frac{2(r-1)K}{n} \qquad (2.9.16)$$

If $V_n(p,\infty)=C$ we have

$$V_n(p,\eta)=C\prod_{r=1}^{n}\left\{\frac{\eta-j\,sn[2(r-1)K/n]}{\eta-j\,sn[2(r-q-1)K/n]}\right\}$$

$$=C\prod_{r=1}^{n}\left[\frac{\eta-j\,sn(2rK/n)}{\eta+j\,sn(2rK/n)}\right] \qquad (2.9.17)$$

from which (2.9.13) follows since

$$S_{11}(p,\eta,\xi)=\frac{V_n(p,\eta)}{V_n(p,\xi)}$$

Property 3
For the match case $A=1$ ($\xi=\infty$) we have

$$S_{12}\left(-jm^{-1/2}\,dc\,\frac{K}{n},\eta,\infty\right)=\prod_{r=1}^{n-1}\left[\frac{1-jm^{1/2}\eta\,sn(2rK/n)}{1+jm^{1/2}\eta\,sn(2rK/n)}\right]^{1/2}$$

$$(2.9.18)$$

Proof

From equation (2.5.1), for $A = 1$ ($\xi = \infty$) and evaluating at the first pole of $F_n(\omega)$, i.e. $\omega = -m^{-1/2}\,\mathrm{dc}(K/n)$, $|S_{12}|^2 = 1$ for η real. Under these conditions, the poles of S_{12} occur when

$$m^{-1/2}\,\mathrm{dc}\frac{K}{n} = \mathrm{cd}\left[\mathrm{sn}^{-1} j\eta + \frac{(2r-1)K}{n}\right] \qquad (2.9.19)$$

Since

$$m^{-1/2}\,\mathrm{dc}\frac{K}{n} = \mathrm{cd}\left[\frac{K}{n} + jK'\right]$$

this reduces to

$$\eta = j\,\mathrm{sn}\left[\frac{2(r-1)K}{n} + jK'\right]$$

$$= jm^{-1/2}\,\mathrm{ns}^{-1}\frac{2(r-1)K}{n} \qquad (2.9.20)$$

from which (2.9.18) follows.

We are now in a position to develop the synthesis procedure to determine the explicit formulas for the element values.

From $S_{11}(p,\eta,\xi)$ defined in equation (2.9.1) we have

$$Z_n(p,\eta,\xi) = \frac{1 + S_{11}(p,\eta,\xi)}{1 - S_{11}(p,\eta,\xi)} \qquad (2.9.21)$$

which is a rational positive function for $\xi > \eta > 0$ and of degree n in p, η and ξ. Furthermore,

$$Z_n(p,\eta,\xi) = -A_n(p,\xi,\eta)$$
$$Z_n(p,\eta,\eta) = \infty \qquad (2.9.22)$$
$$Z_n\left(-j\,\mathrm{cd}\frac{K}{n},\eta,\xi\right) = \infty$$

Completely extracting the pole at $p = -j\,\mathrm{cd}(K/n)$ using a parallel combination of a capacitor and frequency invariant reactance we have

$$Z'_{n-1}(p,\eta,\xi) = Z_n(p,\eta,\xi) - \frac{A_1(\eta,\xi)}{p + j\,\mathrm{cd}(K/n)} \qquad (2.9.23)$$

where $A_1(\eta,\xi)$ is a real rational positive function in η and ξ ($\xi > \eta > 0$). Furthermore,

$$A_1(\eta,\xi) = -A_1(\xi,\eta)$$
$$A_1(\eta,\eta) = \infty \qquad (2.9.24)$$

and

$$Z'_{n-1}(p,\eta,\xi) = -Z'_{n-1}(p,\xi,\eta)$$
$$Z'_{n-1}(p,\eta,\eta) = \infty$$
(2.9.25)

where $Z'_{n-1}(p,\eta,\xi)$ is a positive function.

A frequency invariant reactance $jX_1(\eta,\xi)$ is now extracted from $Z'_{n-1}(p,\eta,\xi)$ such that the remaining impedance $Z''_n(p,\eta,\xi)$ possesses a zero at $p = -j\,\text{cd}(3K/n)$. Therefore

$$Z''_{n-1}(p,\eta,\xi) = Z'_{n-1}(p,\eta,\xi) - jX_1(\eta,\xi)$$
(2.9.26)

where

$$X_1(\eta,\xi) = -X_1(\xi,\eta)$$
(2.9.27)
$$X_1(\eta,\eta) = \infty$$

and

$$Z''_{n-1}(p,\eta,\xi) = -Z''_{n-1}(p,\xi,\eta)$$
$$Z''_{n-1}(p,\eta,\eta) = \infty$$
$$Z''_{n-1}\left(-j\,\text{cd}\,\frac{3K}{n},\eta,\xi\right) = 0$$
(2.9.28)

where $Z''_{n-1}(p,\eta,\xi)$ is a positive function.

An impedance inverter of characteristic impedance $K_{12}(\eta,\xi)$ is now extracted to give

$$Z_{n-1}(p,\eta,\xi) = \frac{K^2_{12}(\eta,\xi)}{Z''_{n-1}(p,\eta,\xi)}$$
(2.9.29)

At the zeros of $K_{12}(\eta,\xi)$, $Z_{n-1}(p,\eta,\xi)$ vanishes and consequently $Z_n(p,\eta,\xi)$ will be of degree one in p. From property 1, the zeros of $K^2_{12}(\eta,\xi)$ are the zeros of

$$\left[\xi + j\,\text{sn}\left(\text{sn}^{-1}\,j\eta + \frac{2K}{n}\right)\right]\left[\xi + j\,\text{sn}\left(\text{sn}^{-1}\,j\eta - \frac{2K}{n}\right)\right]$$
(2.9.30)

and therefore by deliberately choosing $K_{12}(\eta,\eta) = \infty$ we have

$$K_{12}(\eta,\xi) = \frac{\sqrt{\xi^2[1 + \eta^2 m\,\text{sn}^2(2K/n)] - 2\eta\xi\,\text{cn}(2K/n)\text{dn}(2K/n) + \eta^2 + \text{sn}^2(2K/n)}}{\xi - \eta}$$
(2.9.31)

Consequently $Z_{n-1}(p,\eta,\xi)$ is of degree $n-1$ in p, rational in η and ξ

and

$$Z_{n-1}(p,\eta,\xi) = -Z_{n-1}(p,\xi,\eta)$$
$$Z_{n-1}(p,\eta,\eta) = \infty$$
$$Z_{n-1}\left(-j\,cd\,\frac{3K}{n},\eta,\xi\right) = \infty \qquad (2.9.32)$$

with $Z_{n-1}(p,\eta,\xi)$ a positive function for $\xi > \eta > 0$.

Immediately proceeding to the rth cycle in the synthesis procedure we have

$$Z_{n+1-r}(p,\eta,\xi) = \frac{A_r(\eta,\xi)}{p + j\,cd[(2r-1)K/n]} + jX_r(\eta,\xi) + \frac{K^2_{r,r+1}(\eta,\xi)}{Z_{n-r}(p,\eta,\xi)} \qquad (2.9.33)$$

where $A_r(\eta,\xi)$ and $X_r(\eta,\xi)$ are rational functions

$$A_r(\eta,\xi) = -A_r(\xi,\eta),\ X_r(\eta,\xi) = -X_r(\xi,\eta) \qquad (2.9.34)$$
$$A_r(\eta,\eta) = X_r(\eta,\eta) = \infty$$

The zeros of $K_{r,r+1}(\eta,\xi)$ are chosen such that the $Z_n(p,\eta,\xi)$ would be of degree r in p at these points. Thus from property 1, these zeros are zeros of

$$\left[\xi + j\,sn\left(sn^{-1}\,j\eta + \frac{2rK}{n}\right)\right]\left[\xi + j\,sn\left(sn^{-1}\,j\eta - \frac{2rK}{n}\right)\right] \qquad (2.9.35)$$

yielding

$$K_{r,r+1}(\eta,\xi) = \frac{\sqrt{\xi^2[1 + \eta^2 m\,sn^2(2rK/n)] - 2\eta\xi\,cn(2rK/n) + \eta^2 + sn^2(2rK/n)}}{\xi - \eta} \qquad (2.9.36)$$

Furthermore, it readily follows that

$$Z_{n-1}(p,\eta,\xi) = -Z_{n-r}(p,\xi,\eta)$$
$$Z_{n-r}(p,\eta,\eta) = \infty \qquad (2.9.37)$$
$$Z_{n-r}\left(-j\,cd\,\frac{(2r-1)K}{n},\eta,\xi\right) = \infty$$

and $Z_{n-r}(p,\eta,\xi)$ is a positive function for $\xi > \eta > 0$. As in the Chebyshev case, the terminating resistor satisfies the conditions

$$R_L(\eta,\eta) = \infty \quad \text{or} \quad 0 \qquad (2.9.38)$$

Also, from property 3, for $A = 1$ ($\xi = \infty$) the network reduces to a cascade of all-pass sections in $j\eta$ of the form

$$\prod_{r=1}^{n-1} \frac{1}{\sqrt{1 + m\eta^2 \operatorname{sn}^2(2rK/n)}} \begin{bmatrix} -\eta m^{1/2} \operatorname{sn}(2rK/n) & j \\ j & -\eta m^{1/2} \operatorname{sn}(2rK/n) \end{bmatrix}$$

(2.9.39)

and each section may be decomposed as

$$\begin{bmatrix} 1 & j\eta m^{1/2} \operatorname{sn}(2rK/n) \\ 0 & 1 \end{bmatrix}$$

$$\begin{bmatrix} 0 & j\sqrt{1 + \eta^2 m \operatorname{sn}^2(2rK/n)} \\ \dfrac{j}{\sqrt{1 + \eta^2 m \operatorname{sn}^2(2rK/n)}} & 0 \end{bmatrix}$$

$$\begin{bmatrix} 1 & jm^{1/2} \eta \operatorname{sn}(2rK/n) \\ 0 & 1 \end{bmatrix}$$

(2.9.40)

to obtain the correct form for the impedance inverter $K_{r,r+1}(\eta,\infty)$. Between the inverters $K_{r-1,r}$ and $K_{r,r+1}$ we have the series impedance

$$j\eta m^{1/2} \left[\operatorname{sn}\frac{2(r-1)K}{n} + \operatorname{sn}\frac{2rK}{n} \right]$$

(2.9.41)

which must be equal to

$$\frac{A_r(\eta,\infty)}{j\{+ \operatorname{cd}[(2r-1)K/n] - m^{-1/2} \operatorname{dc}(K/n)\}} + jX_r(\eta,\infty)$$

(2.9.42)

Therefore

$$\frac{A_r(\eta,\infty)}{\operatorname{cd}[(2r-1)K/n] - m^{1/2} \operatorname{dc}(K/n)} - X_r(\eta,\infty)$$

$$= -\eta m^{1/2} \left[\operatorname{sn}\frac{2(r-1)K}{n} + \operatorname{sn}\frac{2rK}{n} \right]$$

(2.9.43)

Solving the pair of simultaneous equations (2.9.42) and (2.9.43) for

$A_r(\eta,\infty)$ and $X_r(\eta,\infty)$ yields

$$A_r(\eta,\infty) = \frac{\eta\{\operatorname{sn}[2(r-1)K/n] + \operatorname{sn}(2rK/n)\}}{\operatorname{cd}(K/n)\{1 - m \operatorname{cd}^2(K/n) \operatorname{cd}^2[(2r-1)K/n]\}}$$

$$= 2\eta(1-m) \operatorname{sd}\frac{(2r-1)K}{n} \operatorname{nd}\frac{(2r-1)K}{n} \qquad (2.9.44)$$

and

$$X_r(\eta,\infty) = -\eta m \left[\operatorname{sn}\frac{2(r-1)K}{n} + \operatorname{sn}\frac{2rK}{n}\right] \operatorname{cd}\frac{K}{n} \operatorname{cd}\frac{(2r-1)K}{n} \qquad (2.9.45)$$

establishing the result for the matched case.

Following a similar argument to that used in the Chebyshev design we have, for the arbitrary gain case,

$$R_L(\eta,\xi) = \frac{\xi + \eta}{\xi - \eta} \qquad (2.9.46)$$

and

$$A_r(\eta,\xi) = \frac{\xi A_r(\eta,\infty)}{\xi - \eta} \qquad (2.9.47)$$

From the properties of $X_r(\eta,\xi)$ it also follows that

$$X_r(\eta,\xi) = \frac{\eta\xi X_r(\eta,\infty) + c_r}{\xi - \eta} \qquad (2.9.48)$$

and it remains to determine all the c_r. Using Property 2 we have,

$$S_{11}\left(-j \operatorname{cd}\frac{(2n-1)K}{n}, j\lambda, 0\right) = \prod_{r=1}^{n-1}\left[\frac{\operatorname{sn}(2rK/n) - \lambda}{\operatorname{sn}(2rK/n) + \lambda}\right] \qquad (2.9.49)$$

which is a bounded real unitary reflection coefficient in λ. In the prototype network, for $p = -j \operatorname{cd}[(2n-1)K/n]$, $\eta = j\lambda$, the characteristic impedance of the rth impedance inverter becomes,

$$K_{r,r+1} = \frac{\sqrt{\lambda^2 - \operatorname{sn}^2(2rK/n)}}{\lambda} \qquad (2.9.50)$$

and the impedance of the rth series element is

$$\frac{c_r}{\lambda} \qquad (2.9.51)$$

However, since the reflection coefficient (2.9.19) vanishes at the transmission zero produced by each impedance inverter, from passive all-pass network theory we may uniquely identify (2.9.49) with the reflection coefficient of an open-circuited cascade of all-pass sections with an overall transfer matrix

$$\prod_{1}^{n-1} \left[\frac{1}{\sqrt{\operatorname{sn}^2(2rK/n) - \lambda^2}} \begin{bmatrix} \operatorname{sn}(2rK/n) & \lambda \\ \lambda & \operatorname{sn}(2rK/n) \end{bmatrix} \right] \quad (2.9.52)$$

where each section decomposes into

$$\begin{bmatrix} 1 & \frac{\operatorname{sn}(2rK/n)}{\lambda} \\ 0 & 1 \end{bmatrix} \begin{bmatrix} 0 & 0 \\ \frac{j\lambda}{\sqrt{\lambda^2 - \operatorname{sn}^2(2rK/n)}} & 0 \end{bmatrix} \begin{bmatrix} \frac{j\sqrt{\lambda^2 - \operatorname{sn}^2(2rK/n)}}{\lambda} & 0 \\ 0 & 0 \end{bmatrix}$$

$$\begin{bmatrix} 1 & \frac{\operatorname{sn}(2rK/n)}{\lambda} \\ 0 & 1 \end{bmatrix} \quad (2.9.53)$$

Consequently, equating the series elements, we have

$$c_r = \operatorname{sn} \frac{2(r-1)K}{n} + \operatorname{sn} \frac{2rK}{n} \quad (2.9.54)$$

which completes the solution for the explicit formulas where S_{11} possesses a minimum phase factorization. For alternating zeros of S_{11} which are zeros of

$$1 + j\epsilon\sqrt{1-A} \, \operatorname{cd}_0\left(\frac{nK_0}{K} \operatorname{cd}^{-1} j p\right) \quad (2.9.54)$$

the results are given in the following section and for a proof the reader is referred to Reference 2.4.

2.10 SUMMARY OF RESULTS FOR ELLIPTIC FUNCTION AND INVERSE CHEBYSHEV FILTERS

Referring to Figure 2.10.1 we have

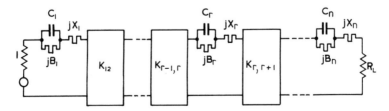

Figure 2.10.1 Prototype for elliptic function and inverse Chebyshev filters

A. The Elliptic Function Prototype

1. Minimum Phase $S_{11}(p)$

$$C_r = \frac{(\xi - \eta)\, ds[(2r-1)K/n]\, dn[(2r-1)K/n]}{2\xi\eta(1-m)}$$

$$B_r = C_r\, cd\, \frac{(2r-1)K}{n} \qquad r = 1 \rightarrow n \qquad (2.10.1)$$

$$X_r = \frac{-\{sn[2(r-1)K/n] + sn(2rK/n)\}\,\{m\xi\eta\, cd[(2r-1)K/n]\, cd(K/n) - 1\}}{\xi - \eta}$$

$$K_{r,r+1} = \frac{\{\xi^2(1+\eta^2 m\, sn^2(2rK/n)] - 2\eta\xi\, cn(2rK/n)\, dn(2rK/n) + \eta^2 + sn^2(2rK/n)\}^{1/2}}{\xi - \eta}$$

$$r = 1 \rightarrow n-1$$

and

$$R_L = \frac{\xi + \eta}{\xi - \eta}$$

where the auxiliary parameters η and ξ are given by

$$\frac{j}{\epsilon} = sn\left[\frac{nK(m_0)}{K(m)} U_0\, |m_0\right], \qquad j\eta = sn(U_0\,|m) \qquad (2.10.2)$$

and

$$\frac{j}{\epsilon\sqrt{1-A}} = sn\left[\frac{nK(m_0)}{K(m)} U_0\, |m_0\right], \qquad j\xi = sn(U_0\,|m)$$

or by using Jacobi's imaginary transformation[2.1]

$$sc\left[\frac{nK(m_0)}{K(m)} U_1\, |1-m_0\right] = \frac{1}{\epsilon}, \qquad \eta = sc(U_1\,|1-m) \qquad (2.10.3)$$

and

$$sc\left[\frac{nK(m_0)}{K(m)} U_1\, |1-m_0\right] = \frac{1}{\epsilon\sqrt{1-A}}, \qquad \xi = sc(U_1\,|1-m)$$

where the elliptic functions in equations (2.10.1) are all dependent upon the elliptic parameter m.

2. Matched Case ($A = 1$, $\xi = \infty$, $R_L = 1$)

$$C_r = \frac{\text{ds}[(2r-1)K/n] \, \text{dn}[(2r-1)K/n]}{2\eta(1-m)} \qquad r = 1 \to n$$

$$B_r = C_r \, \text{cd} \, \frac{(2r-1)K}{n}$$

$$X_r = -\eta m \left[\text{sn} \, \frac{2(r-1)K}{n} + \text{sn} \, \frac{2rK}{n} \right] \text{cd} \, \frac{K}{n} \, \text{cd} \, \frac{(2r-1)K}{n} \qquad (2.10.4)$$

$$K_{r,r+1} = \sqrt{1 + \eta^2 m \, \text{sn}^2 \, \frac{2rK}{n}} \qquad r = 1 \to n-1$$

3. Zeros of $S_{11}(p)$ alternating between the two half-planes

The element values are the same as the matched case apart from a transformer bisecting the central element of turns ratio

$$n = \frac{1 - \sqrt{1-A}}{A^{1/2}} \qquad (2.10.5)$$

4. Zero Impedance Generator ($R_L = 1$, $S_{12} \to Z_{12}$)

$$C_r = \frac{\text{ds}[2r-1)K/n] \, \text{dn}[(2r-1)K/n]}{\eta(1-m)}$$

$$B_r = C_r \, \text{cd} \, \frac{(2r-1)K}{n} \qquad (2.10.6)$$

$$X_r = \frac{-\{\text{sn}[2(r-1)K/n] + \text{sn}(2rK/n)\}\{m\eta^2 \, \text{cd}(K/n) \, \text{cd}[(2r-1)K/n] - 1\}}{2\eta}$$

$$K_{r,r+1} = \frac{\text{sn}(2rK/n)\sqrt{[1 + \eta^2 m \, \text{cd}^2 \, (rK/n)][1 + \eta^2 \, \text{dc}^2 \, (rK/n)]}}{2\eta}$$

B. Inverse Chebyshev Prototype ($m = 0$)

1. Minimum Phase $S_{11}(p)$

$$C_r = \frac{\xi - \eta}{2\xi\eta \sin[(2r-1)\pi/2n]}$$

$$B_r = C_r \cos \frac{(2r-1)\pi}{2n} \qquad r = 1 \to n \qquad (2.10.7)$$

$$X_r = \frac{\sin[(r-1)\pi/n] + \sin(r\pi/n)}{\xi - \eta}$$

$$K_{r,r+1} = \frac{[\xi^2 - 2\eta\xi \cos(r\pi/n) + \eta^2 + \sin^2(r\pi/n)]^{1/2}}{\xi - \eta}$$

$$r = 1 \to n-1$$

and

$$R_L = \frac{\xi + \eta}{\xi - \eta}$$

where the auxiliary parameters η and ξ are given by

$$\eta = \sinh\left(\frac{1}{n} \sinh^{-1} \frac{1}{\epsilon}\right)$$

$$\xi = \sinh\left(\frac{1}{n} \sinh^{-1} \frac{1}{\epsilon\sqrt{1-A}}\right) \qquad (2.10.8)$$

2. Matched Case ($A = 1$, $\xi = \infty$, $R_L = 1$)

$$C_r = \frac{1}{2\eta \sin[(2r-1)\pi/2n]}$$

$$B_r = C_r \cos \frac{(2r-1)\pi}{2n} \qquad r = 1 \to n \qquad (2.10.9)$$

$$X_r = 0$$

$$K_{r,r+1} = 1$$

3. Zeros of $S_{11}(p)$ alternating between two half-planes

Again the only difference from the matched case is that the central element is bisected by the transformer of turns ratio

$$n = \frac{1 - \sqrt{1-A}}{A^{1/2}} \qquad (2.10.10)$$

4. *Zero impedance Generator* $(R_L = 1, S_{12} \to Z_{12})$

$$C_r = \frac{1}{\eta \sin[(2r-1)\pi/2n]}$$

$$B_r = C_r \cos \frac{(2r-1)\pi}{2n} \qquad r = 1 \to n$$

$$X_r = \frac{\sin[(r-1)\pi/n] + \sin(r\pi/n)}{2\eta} \qquad (2.10.11)$$

$$K_{r,r+1} = \frac{\sin(r\pi/2n)\sqrt{\eta^2 + \cos^2(r\pi/2n)}}{\eta} \qquad r = 1 \to n-1$$

One of the most remarkable results which occurs in the recovery of the inverse Chebyshev case is given by equations (2.10.9) for the matched case. In particular $X_r = 0$ resulting in the network being entirely a cascade of resonant elements. Thus, once the first transmission zero has been extracted from the input impedance, the second transmission zero is a zero of the remaining impedance. This property occurs throughout the network if the transmission zeros are extracted in cyclic order.

The results obtained in this section are restricted to the case where the final network possesses quasi-complex conjugate symmetry (exact complex conjugate symmetry in the matched case) and where the entire set of transmission zeros have been extracted in a cyclic manner commencing with the largest negative value. The combination of these two constraints with the minimum phase factorization of the reflection coefficient has led to the element values being linear functions of the auxiliary parameters ξ and η. Consequently the coefficients of these linear functions could be obtained directly from the properties of the three-variable rational elliptic functions without the need to determine an explicit step-by-step synthesis procedure. For realizations incorporating transmission zeros extracted in any other order the element values are no longer linear functions in the auxiliary parameters and could be of degree as high as n^2 in ξ and η. Thus, if explicit formulas do exist in these cases, to obtain them it would be at least necessary to obtain a closed-form expression for the reflection coefficient at each stage in the synthesis procedure.

For the Chebyshev and maximally flat cases, the ordering of the transmission zeros is unique since they are all at infinity and the above comments are not applicable. However, the restriction regarding the minimum (or maximum) phase factorization of the reflection coefficient is applicable to all cases. For an arbitrary choice of left or right half-plane zeros for the numerator of the reflection coefficient, results have recently been obtained for the Chebyshev case but the element values are high-degree functions in the auxiliary parameters[2.14]. The alternating zero case, which has been specifically recovered here as a degeneracy, illustrates this point since if the transformer is transformed to the load, the element values on the output side of the filter are functions of degree $n + 1$ in ξ.

The arbitrary gain case is normally required in problems such as broad-band matching.[2,5] For filtering processes, the matched response ($A = 1$) is normally used or the zero impedance case in some specialized applications. In the following sections we shall be concerned with the matched response.

2.11 DETERMINATION OF THE DEGREE OF THE PROTOTYPE FILTER

Before applying the explicit formulas to the design of practical filters, it is first necessary to determine the degree and ripple levels, if applicable, for the prototype filter in order to meet a required specification. We shall consider each of the four filters in turn with specifications related to the particular response under consideration. Initially we introduce two parameters

$$L_A = 20 \log \frac{1}{S_{12A}}$$

and (2.11.1)

$$L_R = 20 \log \frac{1}{S_{11R}}$$

where L_A is the minimum stopband insertion loss and L_R is the minimum passband return loss; S_{12A} and S_{11R} being the maximum values for the magnitude of S_{12} and S_{11} in the stopband and passband respectively. In addition to L_R and L_A, the ratio of the cut-off frequencies of the passband to stopband is required to determine the degree of the filter. For the maximally flat response, the half-power or 3 dB point is often specified as the cut-off frequency of the passband and this will be considered as a special case. Throughout this section, it will be assumed that the dissipation loss in the filter is zero.

Maximally Flat response

For the maximally flat response

$$|S_{12}(j\omega)|^2 = \frac{1}{1 + \omega^{2n}} \qquad (2.11.2)$$

where the 3 dB point is at $\omega = 1$. Let the cut-off of the required stopband be at $\omega = \omega_s$ and the minimum stopband attenuation be L_A dB. Hence

$$10 \log(1 + \omega_s^{2n}) \geq L_A \qquad (2.11.3)$$

Normally L_A will be sufficiently large to enable unity to be neglected with respect to ω_s^{2n} resulting in the approximate design equation

$$n \geq \frac{L_A}{20 \log \omega_s} \qquad (2.11.4)$$

For example if L_A = 50 dB, ω_s = 2, $n \geq 8.3$, i.e. n = 9.

An alternative method of specification is to give the ratio of stopband to passband cut-off frequencies γ with L_A and L_R. For the stopband requirement equation (2.11.4) is again applicable and for the passband we have

$$10 \log \left[1 + \left(\frac{\gamma}{\omega_s}\right)^{2n}\right] \geq L_R \qquad (2.11.5)$$

Again if L_R is suffiently large we have,

$$n \geq \frac{L_R}{20 \log(\gamma/\omega_s)} \qquad (2.11.6)$$

Combining equations (2.11.5) and (2.11.6) we have the final design equation

$$n \geq \frac{L_A + L_R}{20 \log \gamma} \qquad (2.11.7)$$

For example if L_A = 50 dB, L_R = 20 dB, γ = 2, $n \geq 11.7$, i.e. n = 12.

Chebyshev Response

For the equiripple passband response we have

$$|S_{12}(j\omega)|^2 = \frac{1}{1 + \epsilon^2 T_n^2(\omega)} \qquad (2.11.8)$$

with

$$T_n(\omega) = \cos(n \cos^{-1} \omega)$$
$$= \cosh(n \cosh^{-1} \omega)$$

In the passband ($|\omega| \leq 1$) we have

$$L_R = 10 \log\left(1 + \frac{1}{\epsilon^2}\right)$$

and for most return loss specifications this approximates to

$$L_R = 20 \log \frac{1}{\epsilon} \tag{2.11.9}$$

At the stopband cut-off frequency ω_s, which is equal to γ, the ratio of stopband to passband cut-off frequencies, we have

$$L_A \leqslant 10 \log[1 + \epsilon^2 T_n^2(\gamma)] \tag{2.11.10}$$

Now

$$T_n(\gamma) = \cosh(n \cosh^{-1} \gamma)$$
$$= \cosh[n \ln(\gamma + \sqrt{\gamma^2 - 1})]$$
$$\approx \frac{(\gamma + \sqrt{\gamma^2 - 1})^n}{2} \tag{2.11.11}$$

and therefore for L_A sufficiently large we have

$$L_A \leqslant 20 \left[\log(\epsilon) + \log \frac{(\gamma + \sqrt{\gamma^2 - 1})^n}{2} \right] \tag{2.11.12}$$

Combining equations (2.11.9) and (2.11.12)

$$L_A + L_R \leqslant 20 \log(\gamma + \sqrt{\gamma^2 - 1})^n - 6 \tag{2.11.13}$$

which yields the final design equation

$$n \geqslant \frac{L_A + L_R + 6}{20 \log(\gamma + \sqrt{\gamma^2 - 1})} \tag{2.11.14}$$

Using the same example as given for the maximally flat response, i.e. $L_A = 50$ dB, $L_R = 20$ dB, $\gamma = 2$ we have $n \geqslant 6.6$, i.e. $n = 7$, illustrating the superiority of the Chebyshev design over the maximally flat design for this type of specification.

Inverse Chebyshev Response

For the inverse Chebyshev case we have

$$|S_{12}(j\omega)|^2 = \frac{1}{1 + 1/\epsilon^2 T_n^2(\omega)} \tag{2.11.15}$$

for the high-pass response leading to

$$L_A \approx 20 \log \frac{1}{\epsilon}$$

$$L_R \leqslant 20 \log \frac{\epsilon(\gamma + \sqrt{\gamma^2 - 1})^n}{2} \tag{2.11.16}$$

and the final design equation as in the Chebyshev case is

$$n \geq \frac{L_A + L_R + 6}{20 \log(\gamma + \sqrt{\gamma^2 - 1})} \qquad (2.11.17)$$

Elliptic Function Response

From equations (2.5.1) and (2.5.17) we have for the high-pass response

$$|S_{12}(j\omega)|^2 = \frac{1}{1 + 1/\epsilon^2 F_n^2(\omega)} \qquad (2.11.18)$$

with

$$F_n(\omega) = \mathrm{cd}_0\left(\frac{nK_0}{K}\,\mathrm{cd}^{-1}\,\omega\right) \qquad (2.11.19)$$

the stopband and passband cut-off frequencies being $\omega = 1$ and $m^{-1/2}$ respectively.

From the stopband and passband levels we have

$$L_A = 10 \log\left(1 + \frac{1}{\epsilon^2}\right)$$

$$\approx 20 \log \frac{1}{\epsilon} \qquad (2.11.20)$$

and

$$L_R = 10 \log\left(1 + \frac{\epsilon^2}{m_0}\right)$$

$$\approx 20 \log \frac{\epsilon}{m_0^{1/2}} \qquad (2.11.21)$$

resulting in

$$L_A + L_R = 10 \log \frac{1}{m_0} \qquad (2.11.22)$$

Since m_0 is small,

$$m_0 \approx 16\, e^{-\pi K_0'/K_0}$$

$$= 16\, e^{-\pi n K'/K} \qquad (2.11.23)$$

from the conditional requirement (2.5.12). Hence, we have the approximate design equation

$$n \geq \frac{K(m)}{K'(m)} \frac{(L_A + L_R + 12)}{13.65} \qquad (2.11.24)$$

Using the example $L_A = 50$ dB, $L_R = 20$ dB, $\gamma = 2$, i.e. $m = 0.25$, we have $n \geqslant 4.5$, i.e. $n = 5$.

2.12 FREQUENCY TRANSFORMATIONS AND IMPEDANCE SCALING

In most of the prototype designs, the generator impedance has been normalized to unity. For a value of R all the impedances throughout the network are scaled by R, e.g. $L \rightarrow RL$, $C \rightarrow C/R$, etc.

The frequency transformation from low-pass to low-pass with arbitrary cut-off frequency is $p \rightarrow \alpha p$ and all frequency dependent elements must be scaled by α, e.g. $L \rightarrow \alpha L$, $C \rightarrow \alpha C$. For a high-pass response, $p \rightarrow 1/\alpha p$, i.e. an inductor of value L transforms into a capacitor of value $C = \alpha/L$ and a capacitor of value C transforms into an inductor of value $L = \alpha/C$. Conversely, the transformation $p \rightarrow 1/\alpha p$ maps high-pass filters into low-pass filters. Exact techniques are also available for band-pass and band-stop designs.

Band-pass Response

We shall assume that the prototype network has been transformed to a low-pass response with the passband cut-off frequency $|\omega| = 1$. This prototype is to be transformed into a band-pass structure with the passband extending from $\omega = \omega_1$ to $\omega = \omega_2$ as shown in Figure 2.12.1.

The required frequency transformation is

$$p \rightarrow \alpha \left(\frac{p}{\omega_0} + \frac{\omega_0}{p} \right) \quad (2.12.1)$$

which implies that inductors transform to series resonant circuits and capacitors to parallel resonant circuits as shown in Figure 2.12.2.

To determine α and ω_0 in terms of ω_1 and ω_2 we have the pair of

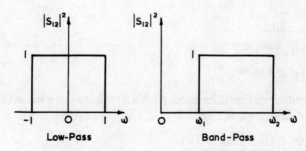

Figure 2.12.1   Band-pass filter specification

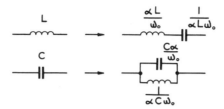

Figure 2.12.2 Low-pass to band-pass transformation of circuit elements

simultaneous equations

$$-1 = \alpha \left(\frac{\omega_1}{\omega_0} - \frac{\omega_0}{\omega_1} \right)$$
$$1 = \alpha \left(\frac{\omega_2}{\omega_0} - \frac{\omega_0}{\omega_2} \right)$$
(2.12.2)

Adding,

$$\frac{\omega_1 + \omega_2}{\omega_0} - \frac{\omega_0}{\omega_1 \omega_2} (\omega_1 + \omega_2) = 0$$

i.e.

$$\omega_0 = \sqrt{\omega_1 \omega_2} \tag{2.12.3}$$

and

$$\alpha = \frac{\sqrt{\omega_1 \omega_2}}{\omega_2 - \omega_1} \tag{2.12.4}$$

In the case of the Chebyshev response, for example, we now have

$$|S_{12}|^2 = \frac{1}{1 + \epsilon^2 T_n^2 [\alpha(\omega/\omega_0 - \omega_0/\omega)]} \tag{2.12.5}$$

but for design purposes it is normally more convenient to transform the band-pass design specification into an equivalent prototype specification to determine required degree etc.

For narrow bandwidths, $\omega_0 \approx \frac{1}{2}(\omega_1 + \omega_2)$ and also α becomes large resulting in impedance values in the filter which are difficult to realize practically. An alternative approximate procedure is presented in the next section.

Band-stop Response

If we assume that the prototype is transformed into a high-pass prototype where the stopband cut-off frequency is normalized to

$|\omega| = 1$, then applying the same transformation as in the band-pass case we obtain a band-stop response with the stopband extending from $\omega = \omega_1$ to $\omega = \omega_2$.

Approximate Frequency Transformations

In the case of the 'natural' prototype for the inverse Chebyshev and elliptic function filters, frequency invariant reactances are present and would be unchanged by the frequency transformation (2.12.1). Consequently, these would remain unrealizable as physical elements and therefore an approximate transformation is required. If a frequency invariant reactance in the network does not contribute in a critical manner to the frequency dependence of that section of the network in which it is located, then it may be replaced by an inductor or capacitor of equal reactance at the band-centre frequency $\omega = \omega_0$. In the vicinity of the passband or stopband $\omega_1 < \omega < \omega_2$, the new element will act approximately equivalently to the desired frequency invariant reactance if the bandwidth is narrow.

For frequency invariant reactances associated with inductors or capacitors in the prototype network, after the band-pass transformation, we equate to an appropriate capacitor—inductor combination. For example, the series combination of an inductor and frequency reactance becomes a series combination of an inductor and capacitor as shown in Figure 2.12.3.

Formally applying the frequency transformation (2.12.1) we have

$$L\alpha \left(\frac{p}{\omega_0} + \frac{\omega_0}{p} \right) + jX \approx L'p + \frac{C'}{p} \qquad (p \approx j\omega_0) \qquad (2.12.6)$$

Equating these functions and their derivatives at $p = j\omega_0$ we have

$$L'\omega_0 = \frac{C'}{\omega_0} = X$$

and

$$L\alpha \left(\frac{1}{\omega_0} + \frac{1}{\omega_0} \right) = L' + \frac{C'}{\omega_0^2} \qquad (2.12.7)$$

Figure 2.12.3 Circuit equivalent of approximate band-pass transformation

yielding

$$L' = \frac{1}{\omega_0}\left(L\alpha + \frac{X}{2}\right)$$

$$C' = \omega_0\left(L\alpha - \frac{X}{2}\right) \tag{2.12.8}$$

This technique is known as the reactance slope parameter method and is valid for narrow bandwidths. This procedure of equating the resonant sections and their derivatives at midband can also be used very successfully to obtain practical impedance values even for very narrow bandwidths. Since in most communication systems channel bandwidths are small, narrow bandwidth design is extremely important and two typical design techniques will be given for specific cases in the following section. It is intended that these examples should provide the basic tools for any engineer designing band-pass or band-stop filters who has to use prescribed types of resonant sections.

2.13 APPROXIMATE DESIGN TECHNIQUES FOR BAND-STOP AND BAND-PASS FILTERS

Band-stop Inverse Chebyshev Filter

We shall assume that we are operating between equal generator and load impedances normalized to 1 Ω. From the high-pass prototype for the matched inverse Chebyshev filter, the unity impedance inverters may be transformed through the network to leave the ladder prototype shown in Figure 2.13.1. The immittance (impedance for series, admittance for shunt) of the rth element is given by (2.10.9)

$$g_r = \frac{2\eta \sin[(2r-1)\pi/2n]}{p + j\cos[(2r-1)\pi/2n]} \tag{2.13.1}$$

where the stopband occurs for $|\omega| \leq 1$. Applying the high-pass to band-stop transformation, equation (2.12.1) gives

$$\frac{1}{g_r} = \frac{\alpha(p/\omega_0 + \omega_0/p) + j\cos\theta_r}{2\eta \sin\theta_r} \qquad \theta_r = \frac{(2r-1)\pi}{2n} \tag{2.13.2}$$

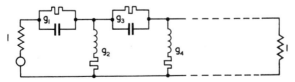

Figure 2.13.1 Ladder prototype for matched inverse Chebyshev filter

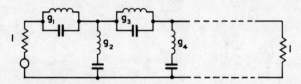

Figure 2.13.2 Narrow-band, band-stop, inverse Chebyshev filter

with

$$\omega_0 = \sqrt{\omega_1 \omega_2}, \quad \alpha = \frac{\sqrt{\omega_1 \omega_2}}{\omega_2 - \omega_1} \qquad (2.13.3)$$

and locates the stopband in $\omega_1 \leq \omega \leq \omega_2$.

Using the approximate transformation (2.12.6), i.e.

$$\frac{1}{g_r} \approx Ap + \frac{B}{p}\bigg|_{p \approx j\omega_0} \qquad (2.13.4)$$

we have

$$A = \frac{2\alpha + \cos \theta_r}{4\omega_0 \eta \sin \theta_r}$$

$$B = \frac{\omega_0 (2\alpha - \cos \theta_r)}{4\eta \sin \theta_r} \qquad (2.13.5)$$

and the final network configuration is shown in Figure 2.3.2.

Thus, we have a band-stop filter, with an equiripple stopband and maximally flat passband constructed entirely from resonant sections.

Narrow Band, Band-pass Chebyshev Filter

Since capacitors may be constructed with low capacitance values, it is possible to realize a band-pass filter, based on a low-pass prototype with all transmission zeros at infinity, using the minimum number of inductors of any desired impedance values. Such a network is illustrated in Figure 2.13.3 and may be designed on an approximate basis over a narrow bandwidth. Initially, consider the π-section of capacitors

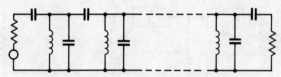

Figure 2.13.3 Narrow-band, band-pass, Chebyshev filter

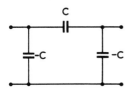

Figure 2.13.4   Approximation to impedance inverter

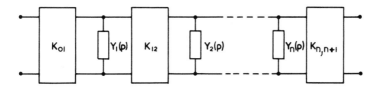

Figure 2.13.5   Network resulting from band-pass transformation

shown in Figure 2.13.4 which for $p = j\omega$ is an ideal impedance inverter of characteristic admittance C (excluding the $1 : -1$ transformer).

Applying the bandpass transformation to the low-pass Chebyshev prototype filter gives the network shown in Figure 2.13.5 where

$$Y_r(p) = C'_r p + \frac{1}{L'_r p} = C_r \alpha \left(\frac{p}{\omega_0} + \frac{\omega_0}{p} \right) \qquad (2.13.6)$$

and

$$C_r = \frac{2}{\eta} \sin \left[\frac{(2r-1)\pi}{2n} \right] \qquad (2.13.7)$$

$$K_{r,\,r+1} = \frac{\sqrt{\eta^2 + \sin^2(r\pi/n)}}{\eta} \qquad (2.13.8)$$

For narrow bandwidths α will be large and the resonant sections would be unrealizable directly. To overcome this problem, the internal impedance level may be scaled by α to give

$$Y_r(p) = \frac{Y_r(p)}{\alpha} \quad \text{and} \quad K'_{r,\,r+1} = \frac{K_{r,\,r+1}}{\alpha} \qquad (2.13.9)$$

However, at the input and output, looking towards the generator and load respectively, we must now see an impedance α. This is achieved around the centre frequency by the circuit shown in Figure 2.13.6.

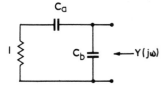

Figure 2.13.6   Circuit for transforming impedance level of network

We require

$$Y(j\omega_0) = jC_b\omega_0 + \frac{1}{1-j/C_a\omega_0}$$

$$= \frac{(C_a\omega_0)^2}{1+(C_a\omega_0)^2} + j\left[C_b\omega_0 + \frac{C_a\omega_0}{1+(C_a\omega_0)^2}\right]$$

$$\approx \frac{1}{\alpha} \tag{2.13.10}$$

Hence,

$$C_a = \frac{1}{\omega_0\sqrt{\alpha-1}} \tag{2.13.11}$$

and

$$C_b = -\frac{\sqrt{\alpha-1}}{\omega_0\alpha} \tag{2.13.12}$$

which will be absorbed into the first resonant section.

After equating the π-sections of capacitors to the inverters we obtain the final circuit shown in Figure 2.13.7 with the design equations

$$C_{0,1} = C_{n,n+1} = \frac{1}{\omega_0\sqrt{\alpha-1}}$$

$$C_{11} = C_{nn} = \frac{C_1}{\omega_0} - \frac{\sqrt{\alpha-1}}{\omega_0\alpha} - \frac{K_{12}}{\alpha}$$

$$C_{r,r+1} = \frac{K_{r,r+1}}{\alpha} \quad r = 1 \to n-1$$

$$C_{r,r} = \frac{C_r}{\omega_0} - \frac{K_{r-1,r}}{\alpha} - \frac{K_{r,r+1}}{\alpha} \quad r = 2 \to n-1$$

$$L_{r,r} = \frac{1}{C_r\omega_0} \quad r = 1 \to n \tag{2.13.13}$$

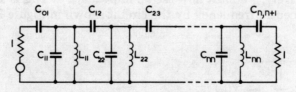

Figure 2.13.7   Final circuit for narrow-band, band-pass, Chebyshev filter

Additional internal scaling may also be applied if the inductors are required to have prescribed values.

This direct-resonator coupled design technique is applicable to most devices which consist of coupled resonator sections including high-frequency components [2.15] such as waveguides.

Modifications to the design equations are necessary for each configuration, but the principle of forming inverters and resonators from which the element values are obtained using the reactance slope parameter technique is the same.

CHAPTER 3

Phase Approximations for Lumped Networks

3.1 INTRODUCTION

For $p = j\omega$ we may write

$$S_{12}(j\omega) = \frac{E_1(\omega) + jO_1(\omega)}{E_2(\omega) + jO_2(\omega)} \tag{3.1.1}$$

where $E_1(\omega)$, $E_2(\omega)$ and $O_1(\omega)$, $O_2(\omega)$ are even and odd polynomials respectively. Thus,

$$S_{12}(j\omega) = |S_{12}(j\omega)| e^{j\psi(\omega)} \tag{3.1.2}$$

where the phase $\psi(\omega)$ is an odd function in ω and given by

$$\psi(\omega) = \tan^{-1}\frac{O_1(\omega)}{E_1(\omega)} - \tan^{-1}\frac{O_2(\omega)}{E_2(\omega)} \tag{3.1.3}$$

If $P_n(p)$ is an nth-degree polynomial in p defined as

$$P_n(j\omega) = [E_2(\omega) + jO_2(\omega)][E_1(\omega) - jO_1(\omega)]$$
$$= E(\omega) + jO(\omega) \tag{3.1.4}$$

then

$$\psi(\omega) = -\tan^{-1}\frac{O(\omega)}{E(\omega)} \tag{3.1.5}$$

In this chapter we shall be concerned with determining the polynomial $P_n(p)$ such that the phase $\psi(\omega)$ approximates to some desired characteristic in a prescribed manner. Initially we shall consider the approximation to an ideal linear phase response in both a maximally flat manner around $\omega = 0$ and also a special finite-band approximation which interpolates to the linear phase characteristics at periodic frequency intervals. Secondly, a maximally flat approximation to a logarithmic phase response will be obtained. We shall then proceed to a finite-band approximation to an arbitrary phase characteristic leading to the introduction of sequences of arbitrary phase polynomials of the first and second kinds.

These various types of approximations will then be shown to have a direct network interpretation for all-pass network functions. These networks are essentially ladder structures and based upon the synthesis of the corresponding reflection filter.

3.2 MAXIMALLY FLAT LINEAR PHASE POLYNOMIAL

For the linear phase response we require that

$$-\psi(\omega) \approx \omega \qquad (3.2.1)$$

over the band of interest where the slope has been normalized to unity. Since $\psi(\omega)$ is an odd function in ω, for an nth-degree polynomial we may equate the error function

$$\epsilon(\omega) = \omega + \psi(\omega) \qquad (3.2.2)$$

and its first $2n$ derivatives to zero at $\omega = 0$ to obtain a maximally flat approximation to the linear phase characteristic. Therefore $\epsilon(\omega)$ may be expanded as

$$\epsilon(\omega) = \epsilon_m(\omega) = a_1 \omega^{2n+1} + a_2 \omega^{2n+3} + \ldots \qquad (3.2.3)$$

where $\epsilon_m(\omega)$ is the error function for the maximally flat approximation.

From equation (3.1.4) we therefore require

$$\frac{P_n(j\omega)}{P_n(-j\omega)} = e^{j2[\omega - \epsilon_m(\omega)]}$$

or

$$P_n(p) e^{-p} - P_n(-p) e^p = b_1 p^{2n+1} + b_2 p^{2n+3} + \ldots \qquad (3.2.4)$$

There are several methods for obtaining the nth-degree polynomial $P_n(p)$ which satisfies equation (3.2.4). One method would be to express $P_n(p)$ as

$$P_n(p) = 1 + a_1 p + a_2 p^2 + \ldots + a_n p^n \qquad (3.2.5)$$

and after substitution in (3.2.4) we obtain the set of linear simultaneous equations

$$\begin{bmatrix} 1 & 0 & 0 & 0 & 0 & \ldots & 0 \\ \frac{1}{2!} & -1 & 1 & 0 & 0 & \ldots & 0 \\ \frac{1}{4!} & -\frac{1}{3!} & \frac{1}{2!} & -1 & 1 & \ldots & 0 \\ \vdots & & & & & & \end{bmatrix} \begin{bmatrix} a_1 \\ a_2 \\ a_3 \\ \vdots \end{bmatrix} = \begin{bmatrix} 1 \\ \frac{1}{3!} \\ \frac{1}{5!} \\ \vdots \end{bmatrix} \qquad (3.2.6)$$

which uniquely determines the coefficients a_r. A second approach is to show that the truncation after n terms of the continued fraction expansion of $\tanh p$ represents the function [3.1]

$$\frac{P_n(p) + P_n(-p)}{P_n(p) - P_n(-p)} \tag{3.2.7}$$

However, we shall adopt the method of representing $P_n(p)$ as a definite integral since it is the only method which allows all of the properties of this polynomial to follow in a direct manner. [3.2] Normalizing the constant term to unity we shall show that $P_n(p)$ may be represented by

$$P_n(p) = \frac{p^{2n+1} e^p}{(2n)!} \int_1^\infty e^{-px}(x^2 - 1)^n \, dx \qquad (\text{Re } p > 0) \tag{3.2.8}$$

To prove that this representation of $P_n(p)$ is correct we first substitute into equation (3.2.4) to give

$$P_n(p) e^{-p} - P_n(-p) e^p = \frac{-p^{2n+1}}{(2n)!} J_n(p) \tag{3.2.9}$$

where

$$J_n(p) = \int_{-1}^1 e^{-px}(x^2 - 1)^n \, dx \tag{3.2.10}$$

and equation (3.2.4) will be satisfied if $J_n(p)$ is an entire function of p (i.e. devoid of poles for all finite values of p). The proof of this condition is relegated to the next section and is recovered as a degenerate case of the finite band equidistant linear phase case.

Secondly we must show that $P_n(p)$ is indeed a polynomial of degree n. This is readily accomplished by deriving a recurrence formula for the generation of $P_n(p)$ as follows.

For $n = 0$

$$P_0(p) = p \, e^p \int_1^\infty e^{-px} \, dx \tag{3.2.11}$$

$$= 1$$

and for $n = 1$

$$P_1(p) = \frac{p^3 e^p}{2} \int_1^\infty e^{-px}(x^2 - 1) \, dx \tag{3.2.12}$$

Defining

$$I_1(p) = \int_1^\infty e^{-px}(x^2 - 1) \, dx \tag{3.2.13}$$

and integrating by parts, we have

$$I_1(p) = \left[\frac{-(x^2-1)e^{-px}}{p}\right]_1^\infty + \frac{2}{p}\int_1^\infty x\,e^{-px}\,dx$$

$$= \frac{2}{p}\left\{\left[\frac{-xe^{-px}}{p}\right]_1^\infty + \left[\frac{-e^{-px}}{p^2}\right]_1^\infty\right\}$$

$$= \frac{2e^{-p}(1+p)}{p^3} \tag{3.2.14}$$

Hence,

$$P_1(p) = 1 + p \tag{3.2.15}$$

Proceeding to the general case for $n \geq 1$ we have

$$P_{n+1}(p) = \frac{p^{2n+3}e^p}{(2[n+1])!}\int_1^\infty e^{-px}(x^2-1)^{n+1}\,dx \tag{3.2.16}$$

Writing

$$I_{n+1}(p) = \int_1^\infty e^{-px}(x^2-1)^{n+1}\,dx \tag{3.2.17}$$

and integrating by parts gives

$$I_{n+1}(p) = \left[\frac{-(x^2-1)^{n+1}e^{-px}}{p}\right]_1^\infty + \frac{2(n+1)}{p}\int_1^\infty e^{-px}x(x^2-1)^n\,dx$$

$$= \frac{2(n+1)}{p}\int_1^\infty e^{-px}x(x^2-1)^n\,dx \tag{3.2.18}$$

for Re $p > 0$ and $n \geq 0$. Integrating by parts for a second time leads to

$$I_{n+1}(p) = \frac{2(n+1)}{p^2}\int_1^\infty e^{-px}[(x^2-1)^n + 2nx^2(x^2-1)^{n-1}]\,dx \tag{3.2.19}$$

for $n \geq 1$, which may be rearranged as

$$I_{n+1}(p) = \frac{2(n+1)}{p^2}\int_1^\infty e^{-px}[(2n+1)(x^2-1)^n + 2n(x^2-1)^{n-1}]\,dx$$

$$= \frac{2(n+1)}{p^2}[(2n+1)I_n(p) + 2nI_{n-1}(p)] \tag{3.2.20}$$

Using equations (3.2.16) and (3.2.17) this recurrence formula reduces to

$$\frac{(2[n+1])!}{p^{2n+3}} P_{n+1}(p) = \frac{2(n+1)(2n+1)(2n)!}{p^{2n+3}} P_n(p)$$

$$+ 4n \frac{(n+1)[2(n-1)]!}{p^{2n+1}} P_{n-1}(p)$$

or

$$P_{n+1}(p) = P_n(p) + \frac{p^2}{(2n+1)(2n-1)} P_{n-1}(p) \tag{3.2.21}$$

Thus, since $P_0(p) = 1$ and $P_1(p) = 1 + p$, $P_n(p)$ is an nth-degree polynomial in p.

Other interesting and useful recurrence formulas may also be derived from the definite integral representation (3.2.8). One such formula relates $P_n'(p)$, the differential of $P_n(p)$, to $P_n(p)$ and $P_{n-1}(p)$. To obtain this we differentiate (3.2.8) to yield

$$P_n'(p) = \frac{p^{2n} e^p}{(2n)!} \{pI_n'(p) + [p + (2n+1)]I_n(p)\} \tag{3.2.22}$$

and

$$I_n'(p) = -\int_1^\infty e^{-px} x(x^2 - 1)^n \, dx \tag{3.2.23}$$

From equation (3.2.18), we have

$$I_n'(p) = -\frac{p}{2(n+1)} I_{n+1}(p) \tag{3.2.24}$$

which, on substitution into (3.2.22), gives

$$P_n'(p) = \frac{p^{2n} e^p}{(2n)!} \left\{ \frac{-p^2}{2(n+1)} I_{n+1}(p) + [p + (2n+1)]I_n(p) \right\}$$

$$= \frac{-(2n+1)}{p} P_{n+1}(p) + \frac{p + (2n+1)}{p} P_n(p)$$

i.e.

$$pP_n'(p) = [p + (2n+1)]P_n(p) - (2n+1)P_{n+1}(p) \tag{3.2.25}$$

Using (3.2.21) this may be reduced to

$$P_n'(p) = P_n(p) - \frac{p}{2n-1} P_{n-1}(p) \tag{3.2.26}$$

Using this recurrence formula it is a relatively simple matter to show directly that the phase of $P_n(p)$ is an optimum maximally flat approximation to the linear phase response by proving that the corresponding group delay is an optimum maximally flat approximation to a constant. In complex form we have the group delay $T_g(p)$ given as

$$T_g(p) = \text{Ev}\left[\frac{P_n'(p)}{P_n(p)}\right] \qquad (3.2.27)$$

$$= 1 - \frac{1}{2n-1}\text{Ev}\left[\frac{pP_{n-1}(p)}{P_n(p)}\right]$$

$$= 1 - \frac{p[P_{n-1}(p)P_n(-p) - P_{n-1}(-p)P_n(p)]}{2(2n-1)P_n(p)P_n(-p)} \qquad (3.2.28)$$

From the degree varying recurrence formula (3.2.21) we have

$$P_{n+1}(p)P_n(-p) - P_{n+1}(-p)P_n(p)$$

$$= \frac{p^2}{(2n+1)(2n-1)}[P_{n-1}(p)P_n(-p) - P_{n-1}(-p)P_n(p)]$$

$$= \frac{(-1)^n 2p^{2n+1}}{(2n+1)\prod_1^n(2r-1)^2}$$

$$= K_{n+1}p^{2n+1} \qquad (3.2.29)$$

Hence, substituting (3.2.29) for $n = n-1$ into (3.2.28) we have

$$T_g(p) = 1 + \frac{A_n^2 p^{2n}}{P_n(p)P_n(-p)} \qquad (3.2.30)$$

where

$$A_n = \frac{1}{1.3.5\ldots(2n-1)} \qquad (3.2.31)$$

demonstrating that $P_n(p)$ is the required polynomial whose phase approximates in a maximally flat manner to a linear phase response at $p = 0$.

One further property of $P_n(p)$ must be investigated if it is to be a useful solution to the approximation problem. From equation (3.1.1) for $S_{12}(p)$ to be a stable function, the zeros of the denominator must be in $\text{Re } p < 0$. In the limiting case where

$$S_{12}(p) = \frac{1}{P_n(p)} \qquad (3.2.32)$$

it is necessary that all of the zeros of $P_n(p)$ be in $\text{Re } p < 0$, i.e. $P_n(p)$ is a strict Hurwitz polynomial. We shall now prove that this is indeed the case by using the recurrence formula (3.2.21) to establish that the sequence of polynomials $P_r(p)$, $r = 0 \to n$, forms a Hurwitz sequence.

Initially we shall construct the impedance function

$$Z_r(p) = \frac{(2r-1)P_r(p)}{pP_{r-1}(p)} \tag{3.2.33}$$

Substitution into (3.2.21) gives

$$Z_{r+1}(p) = \frac{(2r+1)}{p} + \frac{1}{Z_r(p)} \tag{3.2.34}$$

Now if $Z_r(p)$ is a positive real function (p.r.f.) then $1/Z_r(p)$ is a p.r.f. Adding a simple pole at $p = 0$ with a positive real residue implies that $Z_{r+1}(p)$ is a p.r.f. However,

$$Z_1(p) = 1 + \frac{1}{p} \tag{3.2.35}$$

which is a p.r.f. and therefore $Z_r(p)$ is a p.r.f. The network interpretation of this impedance function is shown in Figure 3.2.1 where element values are given as impedance for series arms and admittance for shunt arms and $Z_{r-1}(p)$ is interpreted as an admittance. To maintain the impedance notation for all values of r, we introduce impedance inverters of unity characteristic impedance as shown in Figure 3.2.2.

Since $Z_r(p)$ is a p.r.f. then it must be the ratio of two distinct Hurwitz polynomials of maximum degree r. However, since $P_n(p)$ is of exact degree n, $P_n(p)$ must be a Hurwitz polynomial in p. Furthermore, if $P_n(p)$ had contained zeros on $p = j\omega$, then the delay function $T_g(p)$ would be of degree less than $2n$, which is clearly impossible from

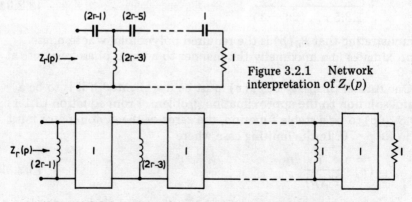

Figure 3.2.1 Network interpretation of $Z_r(p)$

Figure 3.2.2 Network equivalence using impedance inverters

equation (3.2.30), and therefore $P_n(p)$ is a strict Hurwitz polynomial of degree n.

To summarize, we have established that the polynomial $P_n(p)$ whose phase approximates to a linear phase response in an optimum maximally flat manner around $p = 0$, may be readily generated through the recurrence formula

$$P_{n+1}(p) = P_n(p) + \frac{p^2}{(4n^2 - 1)} P_{n-1}(p) \qquad (3.2.36)$$

with the initial conditions

$$P_0(p) = 1, \qquad P_1(p) = 1 + p \qquad (3.2.37)$$

and is a strict Hurwitz polynomial.

3.3 EQUIDISTANT LINEAR PHASE POLYNOMIAL

We shall now proceed to the analytical solution of a polynomial whose phase interpolates to the ideal linear phase response over a finite band about $p = 0$. In particular we shall determine the polynomial which possesses the property

$$P_n(j\omega \mid \epsilon) = A(\omega) e^{j\psi(\omega)} \qquad (3.3.1)$$

where

$$\epsilon(\omega) = [\omega - \psi(\omega)]|_{\omega = r\epsilon} = 0 \qquad r = 0 \rightarrow n$$

i.e. the phase is exactly linear at equidistant frequency increments. This is illustrated in Figure 3.3.1. Following a similar argument to that employed in the maximally flat case we require that

$$P_n(p \mid \epsilon) e^{-p} - P_n(-p \mid \epsilon) e^p = p \prod_{r=1}^{n} [p^2 + (r\epsilon)^2] A_n(p) \qquad (3.3.2)$$

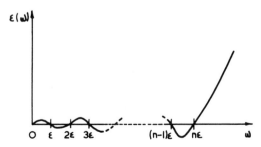

Figure 3.3.1 Equidistant interpolation to linear phase

where $A_n(p)$ is an entire function in p, noting that in the limit

$$P_n(p \mid 0) = P_n(p) \tag{3.3.3}$$

which is the maximally flat solution having adopted the same normalization, i.e. $P_n(0 \mid \epsilon) = 1$. In this particular case we shall prove that $P_n(p \mid \epsilon)$ has a definite integral representation

$$P_n(p \mid \epsilon) = \frac{2^n e^p p \prod_{r=1}^{n} [p^2 + (r\epsilon)^2]}{(2n)! \, \epsilon^{2n} \cos^n \epsilon} \int_1^\infty e^{-px}(\cos \epsilon - \cos \epsilon x)^n \, dx$$

$$\operatorname{Re} p > 0 \tag{3.3.4}$$

Substituting into equation (3.3.2) we have

$$P_n(p \mid \epsilon)e^{-p} - P_n(-p \mid \epsilon)e^p = \frac{-2^n p \prod_{r=1}^{n} [p^2 + (r\epsilon)^2]}{(2n)! \, \epsilon^{2n} \cos^n \epsilon} J_n(p \mid \epsilon)$$

$$\tag{3.3.5}$$

where

$$J_n(p \mid \epsilon) = \int_{-1}^{1} e^{-px}(\cos \epsilon - \cos \epsilon x)^n \, dx \tag{3.3.6}$$

and it must be shown that $J_n(p \mid \epsilon)$ is an entire function in p. Since the kernel of the integral for $J_n(p \mid \epsilon)$ is e^{-px} multiplied by an nth-degree polynomial in $\cos \epsilon x$, and is zero for $x = \pm 1$, we may expand this polynomial as a Fourier series to give

$$J_n(p \mid \epsilon) = \sum_{m=1}^{n} a_m \int_{-1}^{1} e^{-px}(\cos m\epsilon - \cos m\epsilon x) \, dx \tag{3.3.7}$$

and it is only necessary to show that a typical term of the form

$$J_b = \int_{-1}^{1} e^{-px}(\cos b - \cos bx) \, dx \qquad (b = m\epsilon) \tag{3.3.8}$$

is an entire function. Direct integration of (3.3.8) yields

$$J_b = \left[\frac{-e^{-px}}{p} \cos b - \frac{e^{-px}}{p^2 + b^2} (-p \cos bx + b \sin bx) \right]_{-1}^{1}$$

$$= \cos b \, \frac{\sinh p}{2p} - \frac{p \cos b \sinh p + b \sin b \cosh p}{p^2 + b^2} \tag{3.3.9}$$

which, by inspection, is devoid of poles for all finite values of p. Hence $J_n(p \mid \epsilon)$ is an entire function in p.

To prove that $P_n(p \mid \epsilon)$ defined by (3.3.4) is an nth-degree polynomial, we again derive a recurrence formula of the degree varying type. For $n = 0$

$$P_0(p \mid \epsilon) = e^p \, p \int_1^\infty e^{-px} \, dx$$

$$= 1 \qquad (3.3.10)$$

For $n = 1$

$$P_1(p \mid \epsilon) = \frac{2 e^p p(p^2 + \epsilon^2)}{2\epsilon^2 \cos \epsilon} \int_1^\infty e^{-px}(\cos \epsilon - \cos \epsilon x) \, dx$$

$$= \frac{e^p p(p^2 + \epsilon^2)}{\epsilon^2 \cos \epsilon} \left[\frac{-e^{-px}}{p} \cos \epsilon - \frac{e^{-px}}{p^2 + \epsilon^2} (-p \cos \epsilon x + \epsilon \sin \epsilon x) \right]_1^\infty$$

$$= 1 + \frac{\tan \epsilon}{\epsilon} p \qquad (3.3.11)$$

For the general case we may express $P_{n+1}(p \mid \epsilon)$ as

$$P_{n+1}(p \mid \epsilon) = \frac{2^{n+1} e^p p \prod_{r=1}^{n+1} [p^2 + (r\epsilon)^2]}{[2(n+1)]! \, \epsilon^{2(n+1)} \cos^{n+1} \epsilon} I_{n+1}(p \mid \epsilon) \qquad (3.3.12)$$

where

$$I_{n+1}(p \mid \epsilon) = \int_1^\infty e^{-px}(\cos \epsilon - \cos \epsilon x)^{n+1} \, dx \qquad (3.3.13)$$

Integrating by parts,

$$I_{n+1}(p \mid \epsilon) = \left[\frac{-e^{-px}}{p} (\cos \epsilon - \cos \epsilon x)^{n+1} \right]_1^\infty$$

$$+ \int_1^\infty \frac{\epsilon(n+1)}{p} e^{-px} \sin \epsilon x (\cos \epsilon - \cos \epsilon x)^n \, dx$$

$$= \frac{\epsilon(n+1)}{p} \int_1^\infty e^{-px} \sin \epsilon x (\cos \epsilon - \cos \epsilon x)^n \, dx$$

$$(n \geq 0) \quad (3.3.14)$$

Further integration by parts leads to

$$I_{n+1}(p \mid \epsilon) = \frac{\epsilon^2 (n-1)}{p^2} \int_1^\infty e^{-px} [n \sin^2 \epsilon x + \cos \epsilon x (\cos \epsilon - \cos \epsilon x)]$$

$$(\cos \epsilon - \cos \epsilon x)^{n-1} \, dx \qquad n \geq 1 \quad (3.3.15)$$

which, from (3.3.13), may be rearranged to yield

$$I_{n+1}(p \mid \epsilon) = \frac{\epsilon^2 (n+1)}{p^2}$$
$$[n \sin^2 \epsilon \, I_{n-1}(p \mid \epsilon) + (2n+1) \cos \epsilon \, I_n(p \mid \epsilon) - (n+1) I_{n+1}(p \mid \epsilon)]$$
(3.3.16)

Using (3.3.12), we then have the recurrence formula

$$P_{n+1}(p \mid \epsilon) = P_n(p \mid \epsilon) + \left(\frac{\tan \epsilon}{\epsilon}\right)^2 \frac{(p^2 + (\epsilon n)^2)}{(2n+1)(2n-1)} P_{n-1}(p \mid \epsilon)$$
(3.3.17)

and due to the initial conditions (3.3.10) and (3.3.11), this establishes $P_n(p \mid \epsilon)$ as an nth-degree polynomial in p.

We shall now turn to the problem of establishing the conditions on ϵ which will ensure that $P_n(p \mid \epsilon)$ is a Hurwitz polynomial. Initially we observe that for any nth-degree polynomial the maximum possible phase shift for $-\infty \leqslant \omega \leqslant \infty$ is $n\pi$. Since we are interpolating to the phase points $r\epsilon$ for $r = -n \to +n$, a necessary restriction upon ϵ is $|\epsilon| < \pi/2$. For the sake of convenience, and without loss of generality, we may assume that ϵ is non-negative and therefore the necessary restriction is

$$\epsilon < \frac{\pi}{2} \qquad (3.3.18)$$

Turning to sufficiency, we first define the sequence of functions

$$F_{r+1}(p \mid \epsilon) = \frac{\epsilon}{\tan \epsilon} (2r-1)$$
$$\frac{P_{r+1}(p \mid \epsilon)[p^2 + \epsilon^2 (r-1)^2][p^2 + \epsilon^2 (r-3)^2] \cdots}{P_r(p \mid \epsilon)[p^2 + \epsilon^2 r^2][p^2 + \epsilon^2 (r-2)^2] \cdots} \qquad (3.3.19)$$

where the last multiplying factor p^2 in the numerator for r odd or the denominator for r even, is replaced by p. Using the recurrence formula (3.3.17), we have

$$F_{r+1}(p \mid \epsilon) = \frac{\epsilon(2r+1)[p^2 + \epsilon^2 (r-1)^2][p^2 + \epsilon^2 (r-3)^2] \cdots}{\tan \epsilon [p^2 + \epsilon^2 r^2][p^2 + \epsilon^2 (r-2)^2] \cdots}$$
$$+ \frac{1}{F_r(p/\epsilon)} \qquad (3.3.20)$$

with the initial condition

$$F_1(p \mid \epsilon) = 1 + \frac{\epsilon}{\tan \epsilon \cdot p} \qquad (3.3.21)$$

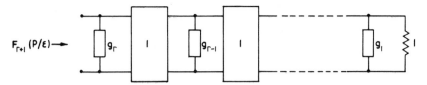

Figure 3.3.2   Network interpretation of $F_{r+1}(p \mid \epsilon)$

If $F_r(p \mid \epsilon)$ is a p.r.f. then $F_{r+1}(p \mid \epsilon)$ is a p.r.f. since the immittance function

$$g_r = \frac{\epsilon(2r+1)[p^2 + \epsilon^2(r-1)^2][p^2 + \epsilon^2(r-3)^2]\cdots}{\tan \epsilon [p^2 + \epsilon^2 r^2][p^2 + \epsilon^2(r-2)^2]\cdots} \quad (3.3.22)$$

is a reactance function for $\epsilon < \pi/2$ due to the interlacing of the poles and zeros on the imaginary axis. This recurrence formula for $F_{r+1}(p \mid \epsilon)$ implies that $F_{r+1}(p \mid \epsilon)$ is the input admittance of the network shown in Figure 3.3.2 where g_r is described by (3.3.22).

Hence, $F_n(p \mid \epsilon)$ is the ratio of two Hurwitz polynomials, and, following a similar argument to the maximally flat case and taking into account the poles and zeros at $p = \pm jr\epsilon$, it may be established that the sequence of polynomials $P_n(p \mid \epsilon)$ are strictly Hurwitz for $\epsilon < \pi/2$. Thus condition (3.3.18) is both necessary and sufficient.

A further useful property of $P_n(p \mid \epsilon)$ which will be exploited in the next chapter may also be derived. From the recurrence formula (3.3.17) we have

$$P_{n+1}(p \mid \epsilon) P_n(-p \mid \epsilon) - P_{n+1}(-p \mid \epsilon) P_n(p \mid \epsilon)$$

$$= \left(\frac{\tan \epsilon}{\epsilon}\right)^2 \frac{p^2 + (\epsilon n)^2}{(2n+1)(2n-1)}$$

$$[P_{n-1}(p \mid \epsilon) P_n(-p \mid \epsilon) - P_{n-1}(-p \mid \epsilon) P_n(p \mid \epsilon)]$$

$$= K_{n+1} p \prod_{r=1}^{n} [p^2 + (\epsilon r)^2] \quad (3.3.23)$$

where

$$K_{n+1} = \left(\frac{\tan \epsilon}{\epsilon}\right)^{2n+1} \frac{1}{(2n+1) \prod_{r=1}^{n} (2r-1)^2} \quad (3.3.24)$$

This equidistant linear phase polynomial will be used in the development of odd-degree transfer functions which approximate simultaneously to an ideal amplitude and linear phase response. For even-degree networks, a modified version is required which arises from the consideration of a polynomial whose phase delay interpolates to a constant value at equidistant frequency intervals.

3.4 EQUIDISTANT CONSTANT PHASE DELAY POLYNOMIAL

Defining the polynomial $Q_n(p \mid \beta)$ ($\beta = \tan \epsilon$) as the polynomial whose phase delay interpolates to a constant in a periodic manner around $p = 0$, along $p = j\omega$, we have

$$Q_n(j\omega \mid \beta) = A(\omega) \, e^{j\psi(\omega)} \tag{3.4.1}$$

where the phase delay $\psi(\omega)/\omega$ possesses the property

$$\frac{\psi(\omega)}{\omega} - \frac{\epsilon}{\beta} = \prod_{r=1}^{n} [p^2 + \beta^2(2r-1)^2] \, B_n(p) \tag{3.4.2}$$

In this case, the points of interpolation are $p = \pm j\beta, \pm j3\beta, \pm j5\beta$, etc., since the origin is no longer such a point. A typical sketch is shown in Figure 3.4.1 for this type of approximation.

To determine this polynomial we require the solution to the equation

$$Q_n(p \mid \beta) \, e^{-\epsilon p/\beta} - Q_n(-p \mid \beta) \, e^{\epsilon p/\beta} = p \prod_{r=1}^{n} [p^2 + \epsilon^2(2r-1)^2] \, A_n(p) \tag{3.4.3}$$

where $A_n(p)$ is an entire function. Again we may write the solution in definite-integral form as

$$Q_n(p \mid \beta) = \frac{\epsilon \cos \epsilon \, e^{\epsilon p/\beta} \prod_{r=1}^{n} [p^2 + \beta^2(2r-1)^2] I_n(p \mid \beta)}{\sin^{2n} \epsilon (2n-1)!} \tag{3.4.4}$$

where

$$I_n(p \mid \beta) = \int_1^{\infty} e^{-\epsilon px/\beta} \sin \epsilon x (\sin^2 \epsilon x - \sin^2 \epsilon)^{n-1} \, dx \tag{3.4.5}$$

and

$\beta = \tan \epsilon$

As in the previous case, to prove that $A_n(p)$ is an entire function

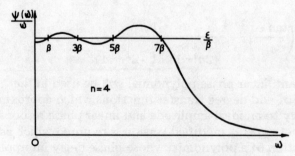

Figure 3.4.1   Equidistant interpolation to constant phase delay

reduces to showing that

$$J_n(p \mid \beta) = \int_{-1}^{1} e^{-\epsilon p x/\beta} \sin \epsilon x (\sin^2 \epsilon x - \sin^2 \epsilon)^{n-1} \, dx \qquad (3.4.6)$$

is an entire function. However, $J_n(p \mid \beta)$ may be expanded as

$$J_n(p \mid \beta) = \sum_{r=1}^{n} a_r \int_{-1}^{1} e^{-\epsilon p x/\beta} \sin[(2r-1)\epsilon x] \, dx \qquad (3.4.7)$$

and the evaluation of a typical term in this series yields

$$J_n(p \mid \beta) = \sum_{r=1}^{n} \frac{2\beta a_r}{\epsilon} \left\{ \frac{(2r-1)\beta \cos[(2r-1)\epsilon] \sinh(\epsilon p/\beta) - p \sin[(2r-1)\epsilon] \cosh(\epsilon p/\beta)}{p^2 + \beta^2 (2r-1)^2} \right\}$$

$$(3.4.8)$$

which, by inspection, is an entire function.

To prove that $Q_n(p \mid \beta)$ is an nth-degree polynomial we again develop a degree-varying recurrence formula.

For $n = 0$ we have the implied condition

$$Q_0(p \mid \beta) = 1 \qquad (3.4.9)$$

and for $n = 1$

$$Q_1(p \mid \beta) = \frac{\epsilon \cos \epsilon \, e^{\epsilon p/\beta} (p^2 + \beta^2)}{\sin^2 \epsilon} \int_1^{\infty} e^{-\epsilon p x/\beta} \sin \epsilon x \, dx$$

$$= 1 + p \qquad (3.4.10)$$

For arbitrary n, consider the integral

$$L_{n+1}(p) = \int_1^{\infty} e^{-px} \sin \epsilon x (\sin^2 \epsilon x - \sin^2 \epsilon)^n \, dx \qquad (3.4.11)$$

where $n > 1$. Integrating by parts gives

$$L_{n+1}(p) = \int_1^{\infty} e^{-px} \frac{\epsilon \cos \epsilon x (\sin^2 \epsilon x - \sin^2 \epsilon)^{n-1} [(2n+1)\sin^2 \epsilon x - \sin^2 \epsilon]}{p} \, dx$$

$$(3.4.12)$$

Performing this operation for a second time yields

$$L_{n+1}(p) = \frac{\epsilon^2}{p^2} \int_1^{\infty} e^{-px} \sin \epsilon x [-(2n+1)^2 (\sin^2 \epsilon x - \sin^2 \epsilon)^n$$

$$+ 2n(2n+1-4\sin^2 \epsilon)(\sin^2 \epsilon x \sin^2 \epsilon)^{n-1}$$

$$+ 4n(n-1)\sin^2 \epsilon(1+\sin^2 \epsilon)(\sin^2 \epsilon x - \sin^2 \epsilon)^{n-2}] \, dx$$

$$(3.4.13)$$

or

$$L_{n+1}(p)[p^2 + \epsilon^2(2n+1)^2] = 2n\epsilon^2(2n+1-4\sin^2\epsilon)L_n(p)$$
$$+ 4n\epsilon^2(n-1)\sin^2\epsilon(1+\sin^2\epsilon)L_{n-1}(p)$$
(3.4.14)

Now since $I_n(p \mid \beta) = L_n(\epsilon p/\beta)$, and using (3.4.4), we obtain the required recurrence formula

$$Q_{n+1}(p \mid \beta) = \left[1 - \frac{(2n-1)\beta^2}{(2n+1)}\right]Q_n(p \mid \beta)$$
$$+ \frac{(p^2 + \beta^2(2n-1)^2)}{(2n+1)(2n-1)}Q_{n-1}(p \mid \beta) \qquad (3.4.15)$$

Using a similar approach to the equidistant linear phase case, $Q_{n+1}(p \mid \beta)$ will be a Hurwitz polynomial if

$$1 - \frac{(2n-1)\beta^2}{(2n+1)} > 0 \qquad (3.4.16)$$

A further useful result occurs if we make the transformation $p \to -p/\beta^2$, $\beta \to 1/\beta$ to give

$$(-1)^n \beta^{2n} p Q_n\left(-\frac{p}{\beta^2} \mid \frac{1}{\beta}\right) = (2n+1)[Q_{n+1}(p \mid \beta) - Q_n(p \mid \beta)] \qquad (3.4.17)$$

since the phase of both sides of this identity match for $p = \pm j\beta(2r-1)$, $r = 1 \to n$, and the highest and lowest coefficients determine the constant multiplier. Defining $P_n(p \mid \beta)$ as

$$Q_n\left(-\frac{p}{\beta^2} \mid \frac{1}{\beta}\right) \cdot (-1)^n \beta^{2n}$$

it follows from (3.4.15) and (3.4.17) that

$$Q_n(p \mid \beta)P_n(-p \mid \beta) + Q_n(-p \mid \beta)P_n(p \mid \beta)$$
$$= K_n \prod_{r=1}^{n} [p^2 + \beta^2(2r-1)^2] \qquad (3.4.18)$$

with the leading coefficient of $P_n(p \mid \beta)$ being the same as $Q_n(p \mid \beta)$ and this is exploited in the next chapter for combined amplitude and phase approximations.

3.5 ARBITRARY PHASE POLYNOMIALS

The next step in an improved approximation technique would be to remove the constraint that the interpolation to a linear phase response should be made at periodic frequency spacings. That is, interpolate to the required linear phase response at

$$p = \pm j\omega_r \qquad r = 1 \to n \tag{3.5.1}$$

where the ω_r are an arbitrary set of prescribed interpolation frequencies. However, the development of the theory requires only minor modifications to allow interpolation to any prescribed phase characteristic, under certain limitations, and therefore this approach is adopted here. Indeed, when processing signals with certain types of modulation, it is advantageous to allow $\psi(\omega)$ to deviate in a prescribed manner from a linear characteristic, particularly towards the passband edge.

We shall define the polynomial $Q_n(p)$ to be the arbitrary phase polynomial of the first kind such that

$$Q_n(\pm j\omega_r) = A(\omega_r) e^{\pm j\psi_r} \tag{3.5.2}$$

where ω_r ($r = 1 \to n$) are the prescribed set of real frequencies, in ascending order of magnitude, and ψ_r are the corresponding prescribed set of interpolated phase values. The solution to the problem defined by (3.5.2) may be obtained as follows.[3.4]
Consider the function

$$F_n(p) = \frac{Q_n(p)}{\prod\limits_{1}^{n}(1 + p^2/\omega_r^2)} \tag{3.5.3}$$

where the ω_r are the prescribed set of interpolating frequencies. The partial fraction expansion of $F_n(p)$ gives

$$F_n(p) = \sum_{1}^{n} \frac{A_r^n(1 + B_r p)}{1 + p^2/\omega_r^2} \tag{3.5.4}$$

and since the denominator of (3.5.3) is an even polynomial we have

$$\text{Arg } Q_n(\pm j\omega_r) = \pm \tan^{-1} \omega_r B_r = \pm \psi_r \tag{3.5.5}$$

where ψ_r is the prescribed value of phase at ω_r. Hence

$$B_r = \frac{\tan \psi_r}{\omega_r} \tag{3.5.6}$$

Since $Q_n(p)$ is constrained to be an nth-degree polynomial in p, when the terms are collected in the numerator of the partial fraction expansion (3.5.4), the first $n - 1$ terms from the highest degree must vanish, i.e. $F_n(p)$ contains an nth-order zero at $p = \infty$. To obtain these conditions, $F_n(p)$ is expanded in a power series about $p = \infty$. From equation (3.5.4),

$$F_n(p) = \sum_{1}^{n} A_r^n \omega_r^2 \left(\frac{1}{p}\right) \left(B_r + \frac{1}{p}\right) \left(1 + \frac{\omega_r^2}{p^2}\right)^{-1}$$

$$= \sum_{1}^{n} A_r^n \omega_r^2 \left(\frac{1}{p}\right) \left[B_r + \frac{1}{p} - B_r \omega_r^2 \left(\frac{1}{p}\right)^2 - \omega_r^2 \left(\frac{1}{p}\right)^3 + \ldots \right]$$

(3.5.7)

Adopting the normalization which restricts the coefficient of p^n in $Q_n(p)$ to be unity, we therefore obtain the following matrix condition

$$[\Omega][A^n] = [I] \quad (3.5.8)$$

where

$$[A^n] = \begin{bmatrix} A_1^n \\ A_2^n \\ \cdot \\ \cdot \\ \cdot \\ A_n^n \end{bmatrix}, [I] = \begin{bmatrix} 0 \\ 0 \\ \cdot \\ \cdot \\ \cdot \\ \gamma \end{bmatrix} \quad (3.5.9)$$

with

$$\gamma = (-1)^{[\frac{1}{2}(n-1)]} \prod_{1}^{n} \omega_r^2 \qquad ([\tfrac{1}{2}(n-1)] = \text{integral part of } \tfrac{1}{2}(n-1))$$

(3.5.10)

and

$$[\Omega] = \begin{bmatrix} \omega_1 \tan \psi_1 & \omega_2 \tan \psi_2 & \omega_3 \tan \psi_3 & \cdots \\ \omega_1^2 & \omega_2^2 & \omega_3^2 & \\ \omega_1^3 \tan \psi_1 & \omega_2^3 \tan \psi_2 & \omega_3^3 \tan \psi_3 & \cdots \\ \omega_1^4 & \cdot & \cdot & \\ \cdot & \cdot & \cdot & \\ \cdot & \cdot & \cdot & \\ \cdot & \cdot & \cdot & \end{bmatrix} \quad (3.5.11)$$

Inverting (3.5.8) we have

$$[A^n] = [\Omega]^{-1}[I] \quad (3.5.12)$$

from which, in principle, the polynomial $Q_n(p)$ may be determined. However, a simpler technique is possible in terms of generating the polynomial through a recurrence formula. Before proceeding to this, it is advantageous to introduce the arbitrary phase polynomial of the second kind defined as $P_n(p)$ such that

$$\text{Arg } P_n(\pm j\omega_r) = \pm \left[\psi_r \pm \frac{\pi}{2}\right] \quad (3.5.13)$$

Following the same technique as in the case of $Q_n(p)$ we have

$$\frac{P_n(p)}{\prod_{1}^{n}(1+p^2/\omega_r^2)} = \sum_{1}^{n}\frac{A_r'^n(1+B_r'p)}{1+p^2/\omega_r^2} \qquad (3.5.14)$$

where

$$B_r' = -\frac{\cot\psi_r}{\omega_r} \qquad (3.5.15)$$

and

$$[A'^n] = [\Omega']^{-1}[I] \qquad (3.5.16)$$

with

$$[\Omega'] = \begin{bmatrix} -\omega_1\cot\psi_1 & -\omega_2\cot\psi_2 & -\omega_3\cot\psi_3 & \cdots \\ \omega_1^2 & \omega_2^2 & \omega_3^2 & \\ -\omega_1^3\cot\psi_1 & -\omega_2^3\cot\psi_2 & -\omega_3^3\cot\psi_3 & \cdots \\ \omega_1^4 & \cdot & \cdot & \\ \cdot & \cdot & \cdot & \\ \cdot & \cdot & \cdot & \\ \cdot & \cdot & \cdot & \end{bmatrix}$$

(3.5.17)

To develop the recurrence properties of $Q_n(p)$ and $P_n(p)$ we first define sequences of polynomials such that the sequence of polynomials from $Q_1(p)$ to $Q_{n+1}(p)$ are such that $Q_{r+1}(p)$ satisfies the same phase conditions as $Q_r(p)$ at $p = \pm j\omega_m$, $m = 1 \to r$, with the additional constraint that Arg $Q_{r+1}(\pm j\omega_{r+1}) = \pm\psi_{r+1}$. A similar definition is adopted for the sequence $P_r(p)$. Noting that we may always express

$$Q_{n+1}(p) - \alpha_n Q_n(p) = (p^2 + \omega_n^2)\left[\sum_{r=0}^{n-1} C_r Q_r(p)\right] \qquad (3.5.18)$$

where $\alpha_n = Q_{n+1}(j\omega_n)/Q_n(j\omega_n) = $ a real constant, since Arg $Q_{n+1}(j\omega_n) = $ Arg $Q_n(j\omega_n)$, we may proceed to determine the coefficients C_r. At $p = j\omega_1$, the argument of the left-hand side of (3.5.18) is ψ_1 and all the terms on the right-hand side are also ψ_1 except for $Q_0(p)$. Hence $C_0 = 0$. Similarly, evaluating at $p = +j\omega_r$ for $r = 1 \to n - 2$ we have $C_r = 0$, $r = 0 \to n - 2$. To obtain C_{r-1}, we notice that the highest coefficient of $Q_{n+1}(p)$ and $Q_{n-1}(p)$ have been normalized to unity, hence $C_{r-1} = 1$. Thus we have the recurrence formula

$$Q_{n+1}(p) = \alpha_n Q_n(p) + (p^2 + \omega_n^2)Q_{n-1}(p) \qquad (3.5.19)$$

where

$$\alpha_n = \frac{Q_{n+1}(j\omega_n)}{Q_n(j\omega_n)} = \left(1 - \frac{\omega_n^2}{\omega_{n+1}^2}\right)\frac{A_n^{n+1}}{A_n^n} \qquad (3.5.20)$$

with the initial conditions

$$Q_0(p) = 1, \qquad Q_1(p) = p + \frac{\omega_1}{\tan \psi_1}$$

Similarly for $P_n(p)$

$$P_{n+1}(p) = \alpha'_n P_n(p) + (p^2 + \omega_n^2) P_{n-1}(p) \tag{3.5.21}$$

However, once $Q_n(p)$ has been determined using the recurrence formula (2.5.19), $P_n(p)$ may be obtained directly from the cross-recurrence formula,

$$pP_n(p) = (p^2 + \omega_n^2) Q_{n-1}(p) + \beta_n Q_n(p) \tag{3.5.22}$$

which may be established in a similar manner to (3.5.19) by using the relationship (3.5.13). In this case

$$\beta_n = \frac{\omega_n}{\tan \psi_n} \frac{A_n^{'n}}{A_n^n} \tag{3.5.23}$$

Combining equations (3.5.19) and (3.5.22) we also have the useful property

$$Q_n(p) P_n(-p) + Q_n(-p) P_n(p) = -2 \prod_1^n (p^2 + \omega_r^2) \tag{3.5.24}$$

To determine the conditions under which the sequence of polynomials $Q_n(p)$ are Hurwitz we again form the immittance function $F_{n+1}(p)$ where, from (3.5.19),

$$F_{n+1}(p) = \frac{\alpha_n (p^2 + \omega_{n-1}^2)(p^2 + \omega_{n-3}^2) \cdots}{(p^2 + \omega_n^2)(p^2 + \omega_{n-2}^2) \cdots} + \frac{1}{F_n(p)} \tag{3.5.25}$$

and

$$F_{n+1}(p) = \frac{Q_{n+1}(p)(p^2 + \omega_{n-1}^2)(p^2 + \omega_{n-3}^2) \cdots}{Q_n(p)(p^2 + \omega_n^2)(p^2 + \omega_{n-2}^2) \cdots} \tag{3.5.26}$$

with the implication that the factor $p^2 + \omega_0^2$ is replaced by p and the initial condition is

$$F_1(p) = 1 + \frac{\omega_1}{\tan \psi_1 p} \tag{3.5.27}$$

This system of equations defines $F_{n+1}(p)$ as the input impedance of a ladder-type structure terminated in a $1\,\Omega$ resistor, where the immitance g_r of the rth element is

$$g_{r+1} = \frac{\alpha_r (p^2 + \omega_{r-1}^2)(p^2 + \omega_{r-2}^2) \cdots}{(p^2 + \omega_r^2)(p^2 + \omega_{r-3}^2) \cdots} \tag{3.5.28}$$

Thus, the sequence of functions $F_r(p)$, $r = 1 \to n+1$, are positive real functions if $\alpha_r \geq 0$ for $r = 1 \to n$, resulting in the sequence of polynomials $Q_r(p)$, $r = 1 \to n+1$, being Hurwitz if $\alpha_r \geq 0$, $r = 1 \to n$.

3.6 MAXIMALLY FLAT LOGARITHMIC PHASE POLYNOMIAL

For the phase approximations considered in this chapter we have restricted our investigation to low-pass characteristics. In Chapter 2, we found that the application of a band-pass or band-stop transformation preserved the shape of the amplitude characteristic but distorted the frequency response such that geometric symmetry was maintained about band-centre. In the phase approximation case, however, this result no longer holds. For example, if $\psi(\omega)$ is the phase of the maximally flat group delay polynomial, i.e.

$$\omega - \psi(\omega) = a_1 \omega^{2n+1} + a_2 \omega^{2n+3} + \ldots \tag{3.6.1}$$

and we apply the band-pass transformation

$$p \to \alpha \left(\frac{p}{\omega_0} + \frac{\omega_0}{p} \right)$$

then if $\psi'(\omega)$ is the phase of the new $2n$th-degree polynomial then

$$\alpha \left(\frac{\omega}{\omega_0} - \frac{\omega_0}{\omega} \right) - \psi'(\omega) = a_1 \alpha^{2n+1} \left(\frac{\omega}{\omega_0} - \frac{\omega_0}{\omega} \right)^{2n+1} + \ldots \tag{3.6.2}$$

Hence, we have achieved a maximally flat approximation about $\omega = \omega_0$ to the phase response

$$\alpha \left(\frac{\omega}{\omega_0} - \frac{\omega_0}{\omega} \right) \tag{3.6.3}$$

and *not* a linear phase characteristic. Thus, the solution to the band-pass problem cannot be obtained from the low-pass prototype. Of course, the band-pass could be tackled directly by determining the polynomial whose phase $\psi(\omega)$ satisfies the condition

$$\omega - \psi(\omega) = a_1 \omega(\omega^2 - \omega_0^2)^n + a_2 \omega(\omega^2 - \omega_0^2)^{n+1} + \ldots \tag{3.6.4}$$

However, there is no analytical solution to this problem although it may be reduced to an eigenvalue problem.[35]

An alternative approach is to provide dispersion in the low-pass prototype such that when the band-pass transformation is applied an approximation to the desired linear phase characteristic can be made. Obviously this implies that the phase for ω negative must differ in magnitude from the corresponding positive value and therefore the required polynomial must possess complex coefficients.

To create dispersion in the low-pass prototype one possible phase to approximate is a quadratic phase of the form

$$\omega + b\omega^2 \tag{3.6.5}$$

with the corresponding linear group delay characteristic

$$T_g = 1 + 2b\omega \tag{3.6.6}$$

However, again there is no known analytical solution. An alternative is to approximate to a linear group velocity characteristic or hyperbolic group delay given by

$$T_g = \frac{1}{1 + \alpha\omega} \tag{3.6.7}$$

This type of characteristic occurs in nature in several ways and is essential for Doppler invariant pulse compression.[3.6] An analytical solution is available in this case and it is also a good match to the dispersion caused by the band-pass transformation.

Integrating (3.6.7) we have the logarithmic phase characteristic

$$\frac{\ln(1 + \alpha\omega)}{\alpha} \tag{3.6.8}$$

which for $\alpha = 0$ degenerates into the linear phase case. Thus, for a maximally flat approximation we require the polynomial of degree n in p, with complex coefficients, whose phase $\psi(\omega)$ satisfies the condition

$$\frac{\ln(1 + \alpha\omega)}{\alpha} - \psi(\omega) = a_1 \omega^{2n+1} + a_2 \omega^{2n+2} + \ldots \tag{3.6.9}$$

or differentiating

$$\frac{1}{1 + 2j\beta p} - \mathrm{Ev}\left[\frac{P_n'(p)}{P_n(p)}\right] = (-1)^n a_1 p^{2n} + \ldots \tag{3.6.10}$$

where

$$\beta = -\frac{\alpha}{2} \quad \text{and} \quad P_n'(p) = \frac{dP_n(p)}{dp}$$

As in the previous cases, a definite integral representation is possible, and specifically expressing the dependence upon the parameter β we shall prove that

$$P_n^{(\beta)}(p) = \frac{(1 + j2\beta p)^{-j/\beta} p^{2n+1} \prod_{r=1}^{n} [1 + (r\beta)^2]}{(2n)!} I_n^{(\beta)}(p) \tag{3.6.11}$$

where

$$I_n^{(\beta)}(p) = \int_1^{\infty} [1 + j(1 + x)\beta p]^{j/\beta - n - 1} (x^2 - 1)^n \, dx \tag{3.6.12}$$

Initially we shall show that $P_n^{(\beta)}(p)$ is a polynomial of degree n by deriving its generating recurrence formula. From (3.6.12)

$$I_{n+1}^{(\beta)}(p) = \int_1^\infty [1+j(1+x)]^{j/\beta - n - 2}(x^2-1)^{n+1}\,dx \tag{3.6.13}$$

Integrating twice, by parts, for $n > 0$ yields

$$I_{n+1}^{(\beta)}(p) = \frac{2(n+1)}{p^2[1+j(n+1)\beta](1+jn\beta)} \int_1^\infty [(x^2-1)^n$$
$$+ 2nx^2(x^2-1)^{n-1}][1+j(1+x)\beta p]^{j/\beta - n}\,dx \tag{3.6.14}$$

which may be expressed as

$$I_{n+1}^{(\beta)}(p) = \frac{2(n+1)}{p[1+j(n+1)\beta](1+jn\beta)} \Big\{(2n+1)(1+j\beta p)I_n^{(\beta)}(p)$$
$$+ 2nI_{n-1}^{(\beta)}(p) + j\frac{(2n+1)\beta}{2(n+1)}p^2[1+j(n+1)\beta]I_{n+1}^{(\beta)}(p)\Big\}$$
$$\tag{3.6.15}$$

or

$$I_{n+1}^{(\beta)}(p) = \frac{2(n+1)}{p^2[1+(n+1)^2\beta^2]}[(2n+1)(1+j\beta p)I_n^{(\beta)}(p) + 2nI_{n-1}^{(\beta)}(p)]$$
$$\tag{3.6.16}$$

Thus, using (3.6.11) we have

$$P_{n+1}^{(\beta)}(p) = (1+j\beta p)P_n^{(\beta)}(p) + \frac{p^2[1+(n\beta)^2]}{(2n+1)(2n-1)}P_{n-1}^{(\beta)}(p) \tag{3.6.17}$$

and by integrating (3.6.12) directly for $n = 0, 1$ we have the initial conditions

$$P_0^{(\beta)}(p) = 1, \qquad P_1^{(\beta)}(p) = 1 + (1+j\beta)p \tag{3.6.18}$$

where the normalization adopted is $P_n^{(\beta)}(0) = 1$.

In order to prove condition (3.6.10), we must first establish a recurrence formula for the differential of $P_n^{(\beta)}(p)$, i.e. $P_n'^{(\beta)}(p)$. From equation (3.6.12)

$$I_n'^{(\beta)}(p) = -[1+j(n+1)\beta]\int_1^\infty [1+j(1+x)\beta p]^{j/\beta - n - 2}(1+x)(x^2-1)^n\,dx$$

$$= -\frac{1+j(n+1)\beta}{1+j2\beta p}\Big\{I_n^{(\beta)}(p) + \frac{[1-j(n+1)\beta]p}{2(n+1)}I_{n+1}^{(\beta)}(p)\Big\} \tag{3.6.19}$$

and using equation (3.6.16), this reduces to
$$p(1 + j2\beta p)I_n^{'(\beta)}(p) = -2nI_{n-1}^{(\beta)}(p)$$
$$- \{2n + 1 + [1 + j\beta(3n + 2)]p\} I_n^{(\beta)}(p) \qquad (3.6.20)$$

However,
$$P_n^{(\beta)}(p) = \frac{(1 + j2\beta p)^{-j/\beta} p^{2n+1} \prod_{1}^{n} [1 + (r\beta)^2]}{(2n)!}$$
$$\left\{\left[\frac{2n+1}{p} + \frac{2}{1+j2\beta p}\right] I_n^{(\beta)}(p) + I_n^{'(\beta)}(p)\right\} \qquad (3.6.21)$$

Substituting (3.6.21) and (3.6.11) into (3.6.20) yields the desired formula
$$(1 + j2\beta p)P_n^{'(\beta)}(p) = (1 + jn\beta)P_n^{(\beta)}(p) - \frac{1 + (n\beta)^2}{2n - 1} pP_{n-1}^{(\beta)}(p) \qquad (3.6.22)$$

It may be noticed that by allowing β to be equal to zero, the recurrence formulas (3.6.17) and (3.6.22) define the maximally flat linear phase polynomial as described in equations (3.2.21) and (3.2.26).

We are now in a position to prove that $P_n^{(\beta)}(p)$ satisfies condition (3.6.10). The group delay function T_g is given by
$$T_g = E_v \left[\frac{P_n^{'(\beta)}(p)}{P_n^{(\beta)}(p)}\right] \qquad (3.6.23)$$

which, by equation (3.6.20) reduces to
$$T_g = \frac{1}{1 + j2\beta p} E_v \left\{\frac{(1 + jn\beta)P_n^{(\beta)}(p) - \dfrac{1 + (n\beta)^2}{2n - 1} pP_{n-1}^{(\beta)}(p)}{P_n^{(\beta)}(p)}\right\}$$
$$= \frac{1}{1 + j2\beta p} \left\{1 - \frac{1 + (n\beta)^2}{2n - 1} E_v \left[\frac{pP_{n-1}^{(\beta)}(p)}{P_n^{(\beta)}(p)}\right]\right\} \qquad (3.6.24)$$

However,
$$E_v \left[\frac{pP_{n-1}^{(\beta)}(p)}{P_n^{(\beta)}(p)}\right] = \frac{p[P_{n-1}^{(\beta)}(p)P_n^{*(\beta)}(-p) - P_{n-1}^{*(\beta)}(-p)P_n^{(\beta)}(p)]}{2P_n^{(\beta)}(p)P_n^{*(\beta)}(p)} \qquad (3.6.25)$$

where the asterisk denotes the replacement of the complex coefficients by their complex conjugates, and from (3.6.17) and (3.6.18)
$$P_{n-1}^{(\beta)}(p)P_n^{*(\beta)}(-p) - P_{n-1}^{*(\beta)}(-p)P_n^{(\beta)}(p)$$
$$= \frac{2(-1)^n p^{2n-1}}{2n - 1} \prod_{r=1}^{n-1} \left[\frac{1 + (r\beta)^2}{(2r-1)^2}\right] \qquad (3.6.26)$$

Hence

$$T_g = \frac{1}{1+j2\beta p}\left\{1 - \frac{(-1)^n p^{2n} \prod_1^n [r+(r\beta)^2]}{P_n^{(\beta)}(p)P_n^{*(\beta)}(-p)\prod_1^n (2r-1)^2}\right\} \quad (3.6.27)$$

which immediately establishes that $P_n^{(\beta)}(p)$ satisfies the desired condition (3.6.10).

To demonstrate that $P_n^{(\beta)}(p)$ is a Hurwitz polynomial with complex coefficients, we define the admittance function

$$Y_{n+1}(p) = \frac{(2n+1)P_{n+1}^{(\beta)}(p)}{pP_n^{(\beta)}(p)} - j(n+1)\beta \quad (3.6.28)$$

and from (3.6.17)

$$Y_{n+1}(p) = \frac{2n+1}{p} + jn\beta + \frac{1+(n\beta)^2}{Y_n(p)+jn\beta} \quad (3.6.29)$$

Thus, if $Y_n(p)$ is a positive function it follows that $Y_{n+1}(p)$ is a positive function and consequently, since $Y_n(p)$ is a positive function, $P_n^{(\beta)}(p)$ is a Hurwitz polynomial.

It is interesting to determine the network produced from the synthesis procedure implied by (3.6.29) Rearranging this equation we have,

$$Y_{n+1}(p) = \frac{2n+1}{p} + \frac{1+jn\beta Y_n(p)}{Y_n(p)+jn\beta} \quad (3.6.30)$$

For $\beta = 0$, the shunt elements are inductors with admittance $2r+1$ and the coupling elements ideal impedance inverters of unity impedance. In this case, the shunt inductors remain the same but the coupling elements, from (3.6.30), possess transfer matrices of the form

$$\frac{1}{\sqrt{1+(r\beta)^2}} \begin{bmatrix} -r\beta & j \\ j & -r\beta \end{bmatrix} \quad (3.6.31)$$

which may be identified with the lossless passive reciprocal, ideal phase shifter of unity characteristic impedance with a transfer matrix

$$\begin{bmatrix} \cos\theta_r & j\sin\theta_r \\ j\sin\theta_r & \cos\theta_r \end{bmatrix} \quad (3.6.32)$$

where

$$\theta_r = -\cot^{-1} r\beta \quad (3.6.33)$$

and the overall network is shown in Figure 3.6.1.

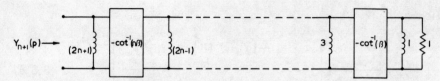

Figure 3.6.1   Network interpretation of logarithmic phase approximation

To convert to a band-pass response, consider the application of the band-pass transformation normalized to a centre frequency $\omega = 1$ given as

$$\omega \to \frac{1}{B}\left(\omega - \frac{1}{\omega}\right) \tag{3.6.34}$$

The group delay of the polynomial

$$H_{2n}(p) = p^n P_n^{(\beta)}\left(\frac{p + 1/p}{B}\right)$$

will then be a maximally flat approximation around $\omega = 1$ to the value

$$T_g = \frac{1}{1 - (2\beta/B)(\omega - 1/\omega)} \frac{1}{B}\left(1 + \frac{1}{\omega^2}\right) \tag{3.6.35}$$

If β is chosen such that

$$\left.\frac{dT_g}{d\omega}\right|_{\omega=0} = 0$$

then

$$\beta = \frac{B}{4} \tag{3.6.36}$$

to give

$$T_g = \frac{2(1 + \omega^2)}{B\omega(1 + 2\omega - \omega^2)} \tag{3.6.37}$$

For a 10% change in ω about $\omega = 1$, the group delay only changes 1% thus illustrating the improvement in the approximation which may be achieved by using this polynomial in preference to the maximally flat linear phase polynomial.

For an arbitrary band-pass transformation using prescribed types of resonators, the required frequency transformation may be approximated by

$$\omega \to F(\omega) \tag{3.6.38}$$

where $F(\omega_0) = 0$; then for

$$\left.\frac{dT_g}{d\omega}\right|_{\omega=\omega_0} = 0$$

we require

$$\beta = \left.\frac{-F''}{2(F')^2}\right|_{\omega=\omega_0} \tag{3.6.39}$$

which will normally imply a relatively broad-band approximation as in the case with (3.6.37). In any network realization, due to the elimination of frequency invariant reactances, further approximations are required but again they are of a broad-band nature.

3.7 ALL-PASS NETWORKS AND REFLECTION FILTERS

The scattering transfer coefficient $S_{12}(p)$ of an all-pass network is defined by

$$S_{12}(p)S_{12}(-p) = 1 \tag{3.7.1}$$

That is, at real frequencies ($p = j\omega$), the amplitude response is unity at all frequencies and provides no amplitude selectivity. If the poles of $S_{12}(p)$ are contained in the Hurwitz polynomial $H_n(p)$ of degree n, then it immediately follows from equation (3.7.1) that

$$S_{12}(p) = \frac{H_n(-p)}{H_n(p)} \tag{3.7.2}$$

and

$$\text{Arg } S_{12}(j\omega) = -2 \text{ Arg } H_n(j\omega) \tag{3.7.3}$$

Thus, if we require an approximation to unity amplitude and linear phase over a prescribed band, we need to determine the polynomial $H_n(p)$ solely from its phase properties. In particular, if we seek a maximally flat approximation about $\omega = 0$ then we require

$$-S_{12}(p) + e^{-2p} = a_1 p^{2n+1} + a_2 p^{2n+2} + \ldots \tag{3.7.4}$$

where the normalization to a delay of 2 at the origin has been chosen purely for convenience. For the maximally flat linear phase polynomial $P_n(p)$ we have, from (3.2.4),

$$e^{-2p} - \frac{P_n(-p)}{P_n(p)} = a_1 p^{2n+1} + a_2 p^{2n+2} + \ldots \tag{3.7.5}$$

and therefore $H_n(p) = P_n(p)$ for the maximally flat solution to the problem. Hence, the required $S_{12}(p)$ is

$$S_{12}(p) = \frac{P_n(-p)}{P_n(p)} \qquad (3.7.6)$$

To synthesize the required network on a transmission basis using a resistively terminated, lossless, reciprocal network the numerator must be either even or odd and consequently augmented to yield

$$S_{12}(p) = \frac{P_n(-p)P_n(p)}{P_n^2(p)} \qquad (3.7.7)$$

and requires a complicated network with complex transmission zeros of degree $2n$. However, if a non-reciprocal realization is sought, a much simpler solution to the synthesis problem is possible. In particular, if $S_{21}(p)$ is chosen to be identically equal to unity then we require a lossless two-port operating between resistive terminations with a two-port scattering matrix

$$\begin{bmatrix} 0 & S_{12}(p) \\ 1 & 0 \end{bmatrix} \qquad (3.7.8)$$

Consider the ideal three-port circulator illustrated in Figure 3.7.1 and defined by the scattering matrix

$$\begin{bmatrix} 0 & e^{j\psi_1} & 0 \\ 0 & 0 & e^{j\psi_2} \\ e^{j\psi_3} & 0 & 0 \end{bmatrix} \qquad (3.7.9)$$

then if port (2) is terminated in a network with a reflection coefficient $S_{11}(p)$, the two-port scattering matrix between ports (1) and (3) becomes, apart from a constant phase shift,

$$\begin{bmatrix} 0 & S_{11}(p) \\ 1 & 0 \end{bmatrix} \qquad (3.7.10)$$

and from (3.7.8) we therefore require a network where $S_{11}(p)$ is equal to the original $S_{12}(p)$. Such a filter is termed a reflection filter.

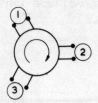

Figure 3.7.1   Ideal three-port circulator

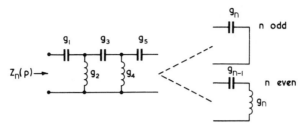

Figure 3.7.2 High-pass ladder network realization of $Z_n(p)$

From (3.7.6) the input impedance of the one port terminating port (2) of the circulator is

$$Z_n(p) = \frac{1 + S_{12}(p)}{1 - S_{12}(p)} = \frac{P_n(p) + P_n(-p)}{P_n(p) - P_n(-p)} \qquad (3.7.11)$$

and since $P_n(p)$ is a Hurwitz polynomial, $Z_n(p)$ is a reactance function. $Z_n(p)$ may be synthesized in numerous ways and one particular method of interest is in the form of a high pass ladder network shown in Figure 3.7.2.

The transfer matrix of such a network may be expressed as

$$[T] = \begin{bmatrix} A\left(\frac{1}{p}\right) & B\left(\frac{1}{p}\right) \\ C\left(\frac{1}{p}\right) & D\left(\frac{1}{p}\right) \end{bmatrix} \qquad (3.7.12)$$

where A, D and B, C are even and odd polynomials respectively,

$$A\left(\frac{1}{p}\right) D\left(\frac{1}{p}\right) - B\left(\frac{1}{p}\right) C\left(\frac{1}{p}\right) = 1 \qquad (3.7.13)$$

and

$$A(0) = D(0) = 1 \qquad (3.7.14)$$

for no transformers.

For n even, the two-port is required to be open-circuited, and from (3.7.4) and (3.7.12) we may make the identification

$$A\left(\frac{1}{p}\right) = \frac{P_n(p) + P_n(-p)}{2a_n p^n}$$
$$C\left(\frac{1}{p}\right) = \frac{P_n(p) - P_n(-p)}{2a_n p^n} \qquad (3.7.15)$$

where a_n is the coefficient of p^n in $P_n(p)$.

Short-circuiting the output we have

$$B\left(\frac{1}{p}\right) = \frac{H_{n-1}(p) + H_{n-1}(-p)}{2b_{n-1}p^{n-1}}$$
$$D\left(\frac{1}{p}\right) = \frac{H_{n-1}(p) - H_{n-1}(-p)}{2b_{n-1}p^{n-1}}$$
(3.7.16)

where $H_{n-1}(p)$ is a polynomial of degree $n-1$ in p and b_{n-1} is the coefficient of p^{n-1}. To determine $H_{n-1}(p)$ we substitute into the reciprocity condition (3.7.13) to yield

$$P_n(-p)H_{n-1}(p) - P_n(p)H_{n-1}(-p) = 2p^{2n-1}a_n b_{n-1} \tag{3.7.17}$$

From (3.2.29), we may therefore immediately identify $H_{n-1}(p)$ with $P_{n-1}(p)$. If $Y(p)$ is the input admittance looking back into the network from the output with a 1 Ω termination on the input side, then

$$Y(p) = \frac{A(1/p) + C(1/p)}{D(1/p) + B(1/p)} \tag{3.7.18}$$

$$= \frac{a_{n-1}P_n(p)}{a_n p P_{n-1}(p)}$$

$$= \frac{(2n-1)P_n(p)}{pP_{n-1}(p)} \tag{3.7.19}$$

which is the same as the immittance function given in equation (3.2.33) which, when synthesized, results in the network shown in Figure 3.2.1. Thus, with reference to Figure 3.7.2 we have the explicit solution

$$g_r = 2r - 1 \tag{3.7.20}$$

Similarly, reflection filters with an all-pass response may be synthesized with other characteristics. For the maximally flat logarithmic phase polynomial the network of Figure 3.6.1 is used where the output is connected to port 2 of the circulator and the input is open-circuited. For the equidistant linear phase polynomial, the resonant ladder structure of Figure 3.3.2 may be used but this results in a non-canonic structure and normally a conventional ladder network would be used.

CHAPTER 4

Simultaneous Amplitude and Phase Approximations for Lumped Networks

4.1 INTRODUCTION

For the solution to the approximation problem for combined amplitude and phase constraints we shall be primarily interested in the low-pass prototype where the transfer function $S_{12}(p)$ approximates to unity gain in the passband and is constrained by the low-pass condition $S_{12}(\infty) = 0$. Another important restriction which will be applied is that the transfer function may be realized directly by a lossless, reciprocal network with resistive terminations. That is

$$S_{12}(p) = \frac{E_{2n-2}(p)}{D_{2n}(p)} \qquad (4.4.1)$$

or

$$S_{12}(p) = \frac{E_{2n}(p)}{D_{2n+1}(p)} \qquad (4.1.2)$$

where $E_{2n-2}(p)$ and $E_{2n}(p)$ are even polynomials and (4.1.1) and (4.1.2) represent even- and odd-degree transfer functions respectively. In addition to this case, we shall also be concerned with transfer functions which may be realized by a non-reciprocal network of the form

$$S_{12}(p) = \frac{H_{n-1}(p)}{D_n(p)} \qquad (4.1.3)$$

particularly with regard to the realization by a reflection filter.

An obvious initial approach to this problem is to attempt to isolate the conditions on amplitude and phase and treat them in an independent manner. There are two distinct possibilities using this approach and they are discussed in the following two sections. However, in one case greater emphasis is placed upon the amplitude response and in the other on the phase response resulting in the necessity to use high-degree transfer functions if compatable constraints on amplitude and phase are required. Thus, the major part of this chapter will be devoted to the case where the constraints cannot be independently considered which results when compatable requirements upon amplitude and phase are required over the same passband.

4.2 CONSTANT AMPLITUDE FILTERS WITH PHASE EQUALIZATION

Consider a transfer function which may be decomposed as

$$S_{12}(p) = S_{12}^A(p) \frac{G(-p)}{G(p)} \qquad (4.2.1)$$

Then

$$|S_{12}(j\omega)| = |S_{12}^A(j\omega)| \qquad (4.2.2)$$

and

$$\text{Arg } S_{12}(j\omega) = \text{Arg } S_{12}^A(j\omega) - 2 \text{ Arg } G(j\omega) \qquad (4.2.3)$$

Thus, the amplitude requirement may be met by the correct choice of $S_{12}^A(p)$ from (4.2.2) and then by constructing the phase response of $S_{12}^A(j\omega)$, $G(p)$ may be determined from (4.2.3) such that the overall phase of $S_{12}(j\omega)$ may be tailored to a desired characteristic. If we consider a maximally flat approximation then

$$|S_{12}^A(j\omega)| = \frac{1}{1 + \omega^{2n}} \qquad (4.2.3)$$

and from Section 2.2 we have

$$S_{12}^A(p) = \frac{1}{\prod_{r=1}^{n}(p - je^{j\theta_r})} \qquad (4.2.4)$$

where

$$\theta_r = \frac{(2r-1)\pi}{2n} \qquad (4.2.5)$$

Let us now consider the possibility of constructing a polynomial $G_m(p)$ of degree m such that a maximally flat approximation of order $2m - 1$ to a linear phase response around $\omega = 0$ may be obtained. To construct the phase response of $S_{12}^A(j\omega)$ we first compile the associated delay, i.e.

$$T_g^A(p) = \text{Ev} \sum_{r=1}^{n} \frac{1}{p - je^{j\theta_r}}$$

$$= \sum_{r=1}^{n} je^{-j\theta_r} \sum_{i=0}^{\infty} (je^{-j\theta_r}p)^{2i} \qquad |p| < 1 \qquad (4.2.6)$$

or at real frequencies

$$T_g^A(j\omega) = \sum_{r=1}^{n} (je^{-j\theta_r} + je^{-j3\theta_r}\omega^2 + je^{-j5\theta_r}\omega^4 + \ldots) \qquad |\omega| < 1$$
$$(4.2.7)$$

A typical term in this series is

$$\sum_{r=1}^{n} je^{-jm\theta_r} \qquad (4.2.8)$$

which may readily be summed to give

$$\sum_{r=1}^{n} je^{-jm(2r-1)\pi/2n} = \frac{j[(-1)^m - 1]}{2\sin(m\pi/2n)} \qquad (4.2.9)$$

Hence,*

$$T_g^A(j\omega) = \frac{1}{\sin(\pi/2n)} + \frac{\omega^2}{\sin(3\pi/2n)} + \frac{\omega^4}{\sin(5\pi/2n)} + \ldots \qquad |\omega| < 1$$
$$(4.2.10)$$

$$= \sum_{i=0}^{\infty} \frac{\omega^{2i}}{\sin[(2i+1)\pi/2n]}$$

To obtain the phase response we integrate to give

$$\psi(\omega) = \sum_{i=0}^{\infty} \frac{\omega^{2i+1}}{(2i+1)\sin[(2i+1)\pi/2n]} \qquad |\omega| < 1 \qquad (4.2.11)$$

If $G_m(p)$ is to be determined to yield a maximally flat linear phase response then

$$\text{Arg } G_m(j\omega) = \sum_{i=0}^{\infty} b_i \omega^{2i+1} \qquad (4.2.12)$$

where

$$b_i = \frac{1}{2(2i+1)\sin[(2i+1)\pi/2n]} \qquad i = 1 \to m \qquad (4.2.13)$$

*Note that when n becomes large this approaches the response given in equation (1.2.17).

However, there is no known analytical solution to determine $G_m(p)$. In the Chebyshev case one may use an interpolation process for which there is an analytical solution. In this case, from Section 2.3,

$$S_{12}^A(p) = \frac{\prod\limits_{1}^{n}[\eta^2 + \sin^2(r\pi/n)]^{1/2}}{\prod\limits_{1}^{n}\{p + j\cos[\sin^{-1} j\eta + (2r-1)\pi/2n]\}} \quad (4.2.14)$$

and using equation (2.7.12) we have

$$\text{Arg } S_{12}^A\left(-j\cos\left[\frac{\pi}{2n}\right]\right) = -\sum_{r=1}^{n-1} \tan^{-1}\frac{\eta}{\sin(\pi/n)} \quad (4.2.15)$$

From (2.10.13) for the degenerate inverse Chebyshev case, by appreciating that $S_{11}(p)$ should replace $S_{12}(p)$ in the Chebyshev case, apart from a constant phase shift we have

$$\text{Arg } S_{12}^A\left(-j\cos\frac{[2q+1]\pi}{2n}\right) = \psi_n + 2\sum_{r=1}^{q} \tan^{-1}\frac{n}{\sin(r\pi/n)} \quad (4.2.16)$$

where ψ_n is the phase at $p = -j\cos(\pi/2n)$.

Hence, combining (4.2.15) and (4.2.16)

$$\text{Arg } S_{12}^A\left(-j\cos\frac{(2q+1)\pi}{2n}\right) = -\sum_{r=1}^{n-1} \tan^{-1}\frac{\eta}{\sin(r\pi/n)}$$

$$+ 2\sum_{r=1}^{q} \tan^{-1}\frac{\eta}{\sin(r\pi/n)}$$

$$q = 0 \to n-1 \quad (4.2.17)$$

Thus, from (4.2.3) we may determine the polynomial of degree $n-1$ which satisfies the conditions,

$$\text{Arg } Q_{n-1}\left(j\cos\frac{(2q+1)\pi}{2n}\right) = K\cos\left[\frac{(2q+1)\pi}{2n}\right]$$

$$+ \tfrac{1}{2}\sum_{r=1}^{n-1}\tan^{-1}\frac{\eta}{\sin(r\pi/n)} - \sum_{r=1}^{q}\tan^{-1}\frac{\eta}{\sin(r\pi/n)}$$

$$q = 0 \to n-1 \quad (4.2.18)$$

where K is a constant and this may be obtained from Section 3.5. A similar result may also be obtained for the elliptic function case.

For this particular Chebyshev solution we have

$$S_{12}(p) = \frac{1}{D_n(p)} \frac{G_{n-1}(-p)}{G_{n-1}(p)} \quad (4.2.19)$$

which for a reciprocal realization must be modified to

$$S_{12}(p) = \frac{1}{D_n(p)} \frac{G_{n-1}(p)G_{n-1}(-p)}{G_{n-1}^2(p)} \tag{4.2.20}$$

which is of degree $3n - 2$. For an arbitrary function of degree $3n - 2$ where the numerator is even, there are $q/2(n-1)$, for n odd, number of arbitrary parameters apart from a bandwidth scaling factor. From (4.2.20) there are $(n-1)$ conditions on passband amplitude, $\frac{1}{2}(n-1)$ on stopband amplitude and $n-1$ on passband phase, yielding a total of $\frac{5}{2}(n-1)$. Thus this particular solution to the combined amplitude and phase constraint is far from optimum since all of the parameters in the transfer function have not been used efficiently. Furthermore, for this reciprocal realization to obtain the linear phase across the passband the degree of the transfer function has increased threefold without any increase in selectivity.

4.3 LINEAR PHASE FILTERS WITH AMPLITUDE EQUALIZATION

If we consider the transfer function given by

$$S_{12}(p) = \frac{1}{P_n(p)} \tag{4.3.1}$$

where $P_n(p)$ is the maximally flat linear phase polynomial. Then $S_{12}(0) = 1$ and the response possesses an optimum maximally flat linear phase response in the passband (around $\omega = 0$) and a maximally flat amplitude in the stopband (around $\omega = \infty$). However the passband amplitude characteristic is poor. To retain the phase response but replace some of the amplitude constraints from the stopband into the passband we may equalize the amplitude by multiplying the numerator by an even polynomial to yield

$$S_{12}(p) = \frac{E_{2m}(p)}{P_n(p)} \tag{4.3.2}$$

where $2m \leq n$. To determine $E_{2m}(p)$ it is useful to notice that since the phase is a maximally flat approximation to a linear response, i.e.

$$S_{12}(j\omega) = A(j\omega)e^{-j(\omega + q_1 \omega^{2n+1} + a_2 \omega^{2n+3} + \ldots)} \tag{4.3.3}$$

then for a maximally flat amplitude of order $2m$ around $\omega = 0$

$$\frac{E_{2m}(p)}{P_n(p)} - e^{-p} = b_1 p^{2m+1} + b_2 p^{2m+2} + \ldots \tag{4.3.4}$$

or $E_{2m}(p)$ is the first $2m$ terms of

$$e^{-p} P_n(p) \tag{4.3.5}$$

Thus, representing $E_{2m}(p)$ as

$$E_{2m}(p) = \sum_{r=0}^{m} a_r p^{2r} \qquad (4.3.6)$$

we immediately have $a_0 = 1$. To obtain a_1 we differentiate w.r.t. p. Expression (4.3.5) becomes

$$e^{-p}[P_n'(p) - P_n(p)] \qquad (4.3.7)$$

which from equation (3.2.26) reduces to

$$\frac{-e^{-p} p P_{n-1}(p)}{2n - 1} \qquad (4.3.8)$$

Hence,

$$A_1 = \frac{-1}{2(2n - 1)} \qquad (4.3.9)$$

Dividing by p and repeating we have

$$a_r = \frac{(-1)^r}{2^r r! \prod_{i=1}^{r} (2n - 2i + 1)} \qquad (r = 1 \to m) \qquad (4.3.10)$$

which is independent of m.

With this particular type of approximation we have the constraint $2m \leq n$ which implies, at most, that only half the conditions which have been applied to the passband phase can be applied to the passband amplitude. Thus, the bandwidth of the linear phase region is always greater than that of the flat amplitude region thus yielding a solution with limited application. Finite-band approximations of this type are also possible but will not be pursued further due to the non-optimum type of solution which results from independently considering the phase and then the amplitude response.

4.4 OPTIMUM MAXIMALLY FLAT CONSTANT AMPLITUDE AND LINEAR PHASE RESPONSE

We shall now consider the cases where the constraints upon the passband amplitude and phase responses cannot be independently considered. For the reciprocal case,

$$S_{12}(p) = \frac{E(p)}{G_n(p)} \qquad (4.4.1)$$

where $E(p)$ is an even polynomial constrained by $S_{12}(\infty) = 0$ and for the maximally flat case by $S_{12}(0) = 1$.

Initially, we apply the maximum number of constraints to the passband amplitude response such that the first $2n-1$ derivatives vanish at $\omega = 0$, i.e.

$$|S_{12}(j\omega)|^2 = \frac{1}{1+K^2\omega^{2n}/[E(j\omega)]^2} \qquad (4.4.2)$$

The remaining $\frac{1}{2}(n-1)$ or $\frac{1}{2}(n-2)$ constraints for n odd or even respectively are applied to the passband phase response in order to force a maximally flat approximation to phase linearity around $\omega = 0$.

Consider the factorization of (4.4.2) as

$$S_{12}(p)S_{12}(-p) = \frac{1}{[1+(-1)^{\frac{1}{2}(n-1)}Kp^n/E(p)][1-(-1)^{\frac{1}{2}(n-1)}Kp^n/E(p)]} \qquad (4.4.3)$$

and for n odd and equal to $2m+1$ we may write

$$S_{12}(p) = \frac{E(p)}{R_1(p)R_2(p)} \qquad (4.4.4)$$

where

$$R_1(p)R_2(-p) = E(p) - Kp^{2m+1} \qquad (4.4.5)$$

or for n even and equal to $2m$,

$$S_{12}(p) = \frac{E(p)}{D_m(p)D_m^*(p)} \qquad (4.4.6)$$

where

$$D_m(p)D_m^*(-p) = E(p) + jKp^{2m} \qquad (4.4.7)$$

$D_m(p)$ is a polynomial with complex coefficients and $D_m^*(p)$ is $D_m(p)$ with the coefficients replaced by their complex conjugates.

For $p = j\omega$ let

$$R_1(j\omega) = A_1(j\omega)e^{j\psi_1(\omega)}$$
$$R_2(j\omega) = A_2(j\omega)E^{j\psi_2(\omega)} \qquad (4.4.8)$$

then from (4.4.4) the phase of $S_{12}(\omega)$ is

$$-\psi(\omega) = \psi_1(\omega) + \psi_2(\omega) \qquad (4.4.9)$$

which is required to be of the form

$$-\psi(\omega) = a_1\omega + a_m\omega^{2m+1} + a_{m+1}\omega^{2m+3} + \ldots \qquad (4.4.10)$$

However, from (4.4.5)

$$\psi_1(\omega) - \psi_2(\omega) = b_m\omega^{2m+1} + b_{m+1}\omega^{2m+3} + \ldots \qquad (4.4.11)$$

Combining the last three conditions results in

$$\psi_1(\omega) = \frac{a_1}{2}\omega + c_m \omega^{2m+1} + c_{m+1}\omega^{2m+3} + \ldots \qquad (4.4.12)$$

and

$$\psi_2(\omega) = \frac{a_1}{2}\omega + d_m \omega^{2m+1} + d_{m+1}\omega^{2m+3} + \ldots \qquad (4.4.13)$$

or that the phase response of the polynomials $R_1(p)$ and $R_2(p)$ must individually be a maximally flat approximation to a linear phase response and both must be of at least degree m. From (4.4.5) since the sum of the degrees is $2m + 1$, then $R_1(p)$ must be of exact degree m and $R_2(p)$ of exact degree $m + 1$. Without loss of generality, we may normalize a_1 to the value 2 and immediately identify

$$R_1(p) = P_m(p) \qquad (4.4.14)$$

where $P_m(p)$ is the maximally flat linear phase polynomial and

$$R_2(p) = (1 + K_1)P_{m+1}(p) - K_1 P_m(p) \qquad (4.4.15)$$

where K_1 is a constant and $P_2(0) = 1$. This constraint may be used either to force an additional constraint on the phase response, i.e. to make a_m zero in equation (4.4.10) or a condition on the stopband amplitude response, i.e. to make $E(p)$ of degree $2m - 2$ rather than degree $2m$.

We shall show, after considering the realization of the associated reflection filter, that in the former case $K = 1$. For the additional stopband constraint from (4.4.5) restricting the degree of $E(p)$ to $2m - 2$ yields $K = (m + 1)/m$. Thus, the odd-degree transfer function with an optimum maximally flat passband amplitude response, a single zero at infinity and $m + 1$ constraints upon a maximally flat linear phase response is

$$S_{12}(p) = \frac{\mathrm{Ev}\{P_m(-p)[2P_{m+1}(p) - P_m(p)]\}}{P_m(p)[2P_{m+1}(p) - P_m(p)]} \qquad (4.4.16)$$

A modified procedure must be employed for the even-degree case. Returning to equation (4.4.3) let $n = 2m$ and then we may write

$$S_{12}(p) = \frac{E(p)}{D_m(p)D_m^*(p)} \qquad (4.4.17)$$

where $D_m(p)$ is a polynomial with complex coefficients and $D_m^*(p)$ is the same polynomial with the coefficients replaced by their complex conjugates. Also,

$$D_m(p)D_m^*(-p) = jE(p) - Kp^{2m} \qquad (4.4.18)$$

For $p = j\omega$ let
$$D_m(j\omega) = A_1(j\omega)e^{j\psi_1(\omega)}$$
then
$$D_m^*(j\omega) = A_1(j\omega)e^{-j\psi_1(-\omega)} \quad (4.4.19)$$
and the phase of the transfer function (4.4.17) is
$$-\psi(\omega) = \psi_1(\omega) - \psi_1(-\omega) \quad (4.4.20)$$
which is required to be of the form
$$-\psi(\omega) = a_1\omega + a_m\omega^{2m+1} + a_{m+1}\omega^{2m+3} + \ldots \quad (4.4.21)$$
However, from (4.4.16),
$$\psi_1(\omega) + \psi_1(-\omega) = \frac{\pi}{2} + b_m\omega^{2m+1} + b_{m+1}\omega^{2m+3} + \ldots \quad (4.4.22)$$
Hence, combining the last three equations,
$$\psi_1(\omega) = \frac{\pi}{4} + \frac{a_1}{2}\omega + c_m\omega^{2m+1} + c_{m+1}\omega^{2m+3} + \ldots \quad (4.4.23)$$
and the most general solution for $D_m(p)$ is
$$D_m(p) = K_1(1+j)P_m(p) + K_2(1-j)pP_{m-1}(p) \quad (4.4.24)$$
Normalizing the constant term in the denominator to be unity results in $K_1 = 1/\sqrt{2}$ and K_2 must be determined from the restriction that $E(p)$ is of maximum degree $2m - 2$. From (4.4.18)
$$E(p) = I_m[D_m(p)D_m^*(-p)]$$
$$= \tfrac{1}{2}P_m(p)P_m(-p) + K_2^2 p^2 P_{m-1}(p)P_{m-1}(-p) \quad (4.4.25)$$
and consequently
$$K_2 = \frac{P_m(p)}{\sqrt{2}pP_{m-1}(p)}\bigg|_{p=\infty}$$
$$= \frac{1}{\sqrt{2}(2m-1)} \quad (4.4.26)$$
using equation (3.2.26). Thus
$$\sqrt{2}D_m(p) = (1+j)P_m(p) + \frac{p}{2m-1}(1-j)P_{m-1}(p) \quad (4.4.27)$$
or alternatively, using (3.2.26),
$$\sqrt{2}D_m(p) = 2P_m(p) - (1-j)P_m'(p) \quad (4.4.28)$$

and

$$S_{12}(p) = \frac{\text{Im.}[D_m(p)D_m^*(-p)]}{D_m(p)D_m^*(p)} \quad (4.4.29)$$

We have not proved that the transfer functions (14.4.16) and (4.4.29) are in fact bounded real functions. However, since $|S_{12}(j\omega)|^2 \leq 1$ due to the constraint (4.4.2) it is only necessary to show that the denominators are Hurwitz polynomials. For (4.4.16) we have shown in Chapter 3 that $P_m(p)$ is Hurwitz. Also, using the degree-varying recurrence formula for $P_m(p)$ (3.2.21) the polynomial

$$2P_{m+1}(p) - P_m(p) = P_m(p) + \frac{2p^2}{4m^2 - 1} P_{m-1}(p)$$

$$= pP_{m-1}(p) \cdot \left[\frac{P_m(p)}{pP_{m-1}(p)} + \frac{2p}{4m^2 - 1} \right] \quad (4.4.30)$$

and since the first term in the bracket is a positive real function from Section 3.2

$$\frac{P_m(p)}{pP_{m-1}(p)} + \frac{2p}{4m^2 - 1} \quad (4.4.31)$$

is also a p.r.f. and consequently all of its zeros are in the left half-plane.

For equation (4.4.29) the polynomial $D_n(p)$ may be written, using (4.4.27), as

$$\sqrt{2} \frac{(1-j)D_m(p)}{P_m(p)} = j + \frac{pP_{m-1}(p)}{(2m-1)P_m(p)} \quad (4.4.32)$$

which is a positive function and consequently all of the zeros are in the left half-plane.

We shall now turn to the realization of these particular transfer functions using reflection filters and derive the explicit formulas for the element values. Using a similar technique to that used in Section 3.7, we require a two-port lossless network with a transfer function of the form

$$|S_{12}(j\omega)|^2 = \frac{K^2 \omega^{2n}}{E^2(j\omega) + K^2 \omega^{2n}} \quad (4.4.33)$$

and

$$|S_{11}(j\omega)|^2 = \frac{E^2(j\omega)}{E^2(j\omega) + K^2 \omega^{2n}} \quad (4.4.34)$$

Thus, from (4.4.33) the network must be a high-pass ladder network and since the numerator of $S_{11}(p)$ is odd or even in $1/p$, the network will be symmetrical or antisymmetric. In this case it is also possible, using the bisection theorem, to write the transfer and reflection coefficients in terms of the even- and odd-mode input admittance functions which are obtained from placing either an open- or a short-circuit along the line of symmetry of the network. If the even-mode admittance is Y_e, and the odd-mode admittance Y_o, then

$$S_{12}(p) = \frac{Y_e - Y_o}{(1 + Y_e)(1 + Y_o)} \qquad (4.4.35)$$

$$S_{11}(p) = \frac{1 - Y_e Y_o}{(1 + Y_e)(1 + Y_o)}$$

where Y_e and Y_o are both reactance functions. From (4.4.16)

$$Y_o = \frac{P_m(p) + P_m(-p)}{P_m(p) - P_m(-p)} \qquad (4.4.36)$$

and

$$Y_e = \frac{\mathrm{Ev}[2P_{m+1}(p) - P_m(p)]}{\mathrm{Od}[2P_{m+1}(p) - P_m(p)]} \qquad (4.4.37)$$

Since we have already established that the network is a high-pass ladder network, we may obtain every element apart from the central one using only Y_o. From Section 3.7, Y_o is the input admittance of the network shown in Figure 4.4.1.

Computing the transfer matrix for this network gives, from (3.7.15) and (3.7.16),

$$[T] = \frac{1}{p^{m-1}} \begin{bmatrix} \dfrac{P_m(p) - P_m(-p)}{a_1 p} & \dfrac{P_{m-1}(p) - P_{m-1}(-p)}{b_1} \\ \dfrac{P_m(p) + P_m(-p)}{a_1 p} & \dfrac{P_{m-1}(p) + P_{m-1}(-p)}{b_1} \end{bmatrix} \qquad (4.4.38)$$

If an impedance inverter followed by a shunt inductor of value L_r are added to the output of the network, the open-circuited input

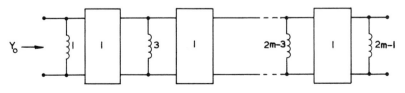

Figure 4.4.1   Network realization of odd-mode admittance

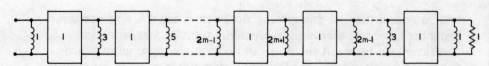

Figure 4.4.2   Final realization of odd-degree reflection filter

admittance must be Y_e, i.e.

$$Y_e = \frac{L_r[P_m(p) + P_m(-p)] + [p^2/(2m-1)][P_{m-1}(p) + P_{m-1}(-p)]}{L_r[P_m(p) - P_m(-p)] + [p^2/(2m-1)][P_{m-1}(p) - P_{m-1}(-p)]} \quad (4.4.39)$$

$$= \frac{\mathrm{E_v}\{[1 - (2m+1)/L_r]P_m(p) + [(2m+1)/L_r]P_{m+1}(p)\}}{\mathrm{Od}\{[1 - (2m+1)/L_r]P_m(p) + [(2m+1)/L_r]P_{m+1}(p)\}} \quad (4.4.40)$$

using the degree-varying recurrence formula for $P_m(p)$, and identifying this expression with (4.4.37) gives

$$L_r = \frac{2m+1}{2} \quad (4.4.41)$$

Thus, the complete network is shown in Figure 4.4.2.

The value of $2m+1$ for the central element is not coincidental. Since in deriving the original formula (4.4.16) we insisted upon the extra parameter being applied to the phase characteristic, the input impedance must approximate to tanh p for one extra term of its continued-fraction expansion.

A similar process can be developed for the even-degree case but Y_e and Y_o now have complex coefficients and $Y_0 = Y_e^*$. The result of synthesizing the reflection filter is shown in Figure 4.4.3 where the central element is now an impedance inverter which, upon forming the even- and odd-mode admittances, decomposes into two 45°, unity impedance, phase shifters.

The equivalent circuits for these reflection filters were derived from the prescribed transfer functions which in turn were obtained from mathematical constraints upon the amplitude and phase. However, using some basic results from passive network theory one may readily

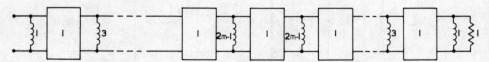

Figure 4.4.3   Final realization of even-degree reflection filter

demonstrate how the equivalent circuits shown in Figures 4.4.2 and 4.4.3 may be obtained directly.

Since the maximum number of maximally flat constraints were placed upon the passband amplitude response all of the transmission zeros of the reflection network must be at the origin, that is it must be a high-pass ladder network. Furthermore, since the numerator of the reflection coefficient is an even polynomial in p, the reflection network must be symmetrical. Finally, we insist upon the delay response being a maximally flat approximation to a constant about the origin with the maximum number of constraints. Thus the network should approximate to the continued-fraction expansion of $\tanh p$ around $p = 0$ for the maximum possible number of terms. Hence, the rth inductance should be of value $(2r-1)$ up to the maximum number of elements within the constraint of a symmetrical network. For an odd-degree network this maximum value is $\frac{1}{2}(n+1)$ and for the even-degree case, $\frac{1}{2}n$. Furthermore, there must be perfect transmission at $p = \infty$.

If one removes the constraint of a symmetrical realization, then the numerator of the reflection coefficient will no longer be even but one may place more constraints upon the delay response. Unfortunately the transfer function of the corresponding transmission filter will no longer be directly realizable in a canonic manner using a reciprocal realization. However, if the transfer function is to be realized using the reflection filter as given in Section 3.7, then the optimum reflection network will be as shown in Figure 4.4.4 where the unity terminating resistor is required to provide the required unity gain at $p = \infty$.

From the transfer matrix we obtain the reflection coefficient

$$S_{11}(p) = \frac{(2n-1)P_n(-p) + pP_{n-1}(-p)}{(2n-1)P_n(p) + pP_{n-1}(p)} \tag{4.4.42}$$

which, for $p = j\omega$, must possess the property

$$S_{11}(j\omega) = A(j\omega)e^{j\psi(\omega)}$$

where

$$A^2(j\omega) = 1 - a_1\omega^{2n} - a_2\omega^{2n+3} \tag{4.4.43}$$

$$-\psi(\omega) = 2\omega + c_1\omega^{2n+1} + c_2\omega^{2n+3} \tag{4.4.44}$$

and

$$S_{11}(\infty) = 0$$

Figure 4.4.4   Reflection filter with maximum number of phase constraints

Before proceeding to finite-band approximations for combined amplitude and phase response, the transfer function which approximates to an ideal amplitude and logarithmic phase response in maximally flat manner will be considered.

4.5 OPTIMUM MAXIMALLY FLAT CONSTANT AMPLITUDE AND LOGARITHMIC PHASE RESPONSE

In Section 3.6 it was shown that when one transformed a low-pass prototype into a band-pass filter, group delay distortion occurs, and that by using a dispersive prototype with a logarithmic phase response good linear phase band-pass responses could be obtained. Since the amplitude constraints are not affected by frequency transformation, it is apparent that it would be desirable to obtain the solution for the transfer function which possesses the property

$$S_{12}(j\omega) \approx (1 - 2\beta\omega)^{j/\beta} \quad |\omega| < \omega_c$$
$$\approx 0 \quad |\omega| > \omega_c \quad (4.5.1)$$

and approximates to this characteristic in an optimum maximally flat manner about $\omega = 0$.

It is of interest to note that this response characteristic is required for the compression of frequency modulated pulses where the modulation is chosen to be invariant to Doppler effects.[3.6] Additionally, it is essential for expansion and compression of signals using surface acoustic wave devices if the temperature difference between the transmitter and receiver filters is unknown and a matched response is to be in independently maintained. In these applications it is possible to use the reflection type of realization and we shall consider this case first.

From Figure 3.6.1, using similar arguments to those put forward in the previous section, we can immediately deduce that the required network is as given in Figure 4.5.1. Using the results of Sections 3.6 and 4.4 it follows that

$$S_{11}(p) = \frac{(2n-1)P_n^{*(\beta)}(-p) + p(1 - jn\beta)P_{n-1}^{*(\beta)}(-p)}{(2n-1)P_n^{(\beta)}(p) + p(1 + jn\beta)P_{n-1}^{(\beta)}(p)} \quad (4.5.2)$$

and possesses the properties

$$|S_{11}(j\omega)|^2 = 1 - a_1\omega^{2n} - a_2\omega^{2n+1} - a_3\omega^{2n+2} \quad (4.5.3)$$

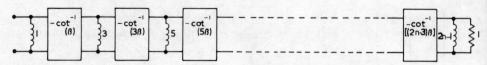

Figure 4.5.1   Reflection filter with logarithmic phase

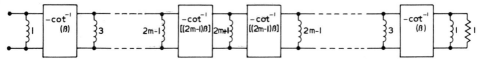

Figure 4.5.2 Reflection filter for corresponding reciprocal realization (n odd)

$$-\text{Arg } S_{11}(j\omega) = \frac{1}{\beta}\ln(1 - 2\beta\omega) + c_1\omega^{2n} + c_2\omega^{2n+1} + \ldots \quad (4.5.4)$$

and

$$S_{11}(\infty) = 0$$

For reciprocal transmission types of realizations, the associated reflection filters must be symmetrical and again we may readily deduce that the corresponding reflection filters must be of the form shown in Figures 4.5.2 and 4.5.3 for n odd and even respectively.

Direct analysis of these networks yields for n odd

$$S_{12}(p) = \frac{E_v[P_n^{*(\beta)}(p)\{2P_{n+1}^{(\beta)}(p) - [1 + j2(n+1)\beta p/(2n+1)]P_n^{(\beta)}(p)\}]}{P_n^{(\beta)}(p)\{2P_{n+1}^{(\beta)}(p) - [1 + j2(n+1)\beta p/(2n+1)]P_n^{(\beta)}(p)\}} \quad (4.5.5)$$

and for n even

$$S_{12}(p) = \frac{\text{Im}[Q_n^{(\beta)}(p)R_n^{*(\beta)}(p)]}{Q_n^{(\beta)}(p)R_n^{(\beta)}(p)} \quad (4.5.6)$$

where

$$Q_n^{(\beta)}(p) = (1+j)[P_n^{(\beta)}(p) - \frac{j\beta np}{2n-1}P_{n-1}^{(\beta)}(p)] + \frac{(1-j)p}{2n-1}P_{n-1}^{(\beta)}(p) \quad (4.5.7)$$

and

$$R_n^{(\beta)}(p) = (1-j)[P_n^{(\beta)}(p) - \frac{j\beta p}{2n-1}P_{n-1}^{(\beta)}(p)] - \frac{(1+j)p}{2n-1}P_{n-1}^{(\beta)}(p) \quad (4.5.8)$$

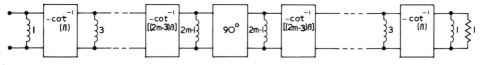

Figure 4.5.3 Reflection filter for corresponding reciprocal realization (n even)

which further reduces to

$$S_{12}(p) = \frac{P_n^{(\beta)}(p)P_n^{*(\beta)}(p) + [(1-\beta^2 n^2)p^2/(2n-1)^2]P_{n-1}^{(\beta)}(p)P_{n-1}^{*(\beta)}(p)}{\{P_n^{(\beta)}(p) - [jnp/(2n-1)]P_{n-1}^{(\beta)}(p)\}^2 + p^2 P_{n-1}^{2(\beta)}(p)/(2n-1)^2} \quad (4.5.9)$$

From equation (4.5.1), the group delay for $|\omega| < \omega_c$ approximates to

$$T_g = \frac{2}{1 - 2\beta\omega} \quad (4.5.10)$$

Applying a band-pass frequency transformation

$$\omega \to \frac{1}{\beta}\left(\omega - \frac{1}{\omega}\right) \quad (4.5.11)$$

results in a group delay

$$T_g = \frac{2}{[1 - (2\beta/B)(\omega - 1/\omega)]} \frac{1}{B}\left(1 + \frac{1}{\omega^2}\right) \quad (4\ 5.12)$$

and β may be chosen as described in Section 3.6 such that $dT_g/d\omega$ at the centre frequency $\omega = 1$ is zero, i.e.

$$\beta = \frac{B}{4} \quad (4.5.13)$$

Thus, a band-pass transfer characteristic has been formed which approximates to unity amplitude in a maximally flat manner around

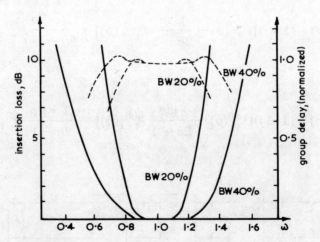

Figure 4.5.4   Computed response for band-pass linear phase filters

$\omega = 1$ and the group delay is approximately maximally flat to a constant value also around $\omega = 1$. However, any direct realization of the prototype such as in the form of a cascade[4.1] will result in frequency invariant reactances. But by using the reactance slope parameter technique as described in Chapter 2, these may be removed after applying the band-pass frequency transformation. Even after using all of these approximations the response still approximates to the ideal even for filters of relatively large fractional bandwidths as may be seen from Figure 4.5.4 for 5th-degree networks with 20% and 40% fractional bandwidths. A further point of interest is the approximate arithmetical symmetry of the response in contrast to the normal geometric symmetry which would have resulted without dispersion in the prototype.

This completes the maximally flat solutions and we now consider the important finite-band approximations for constant amplitude and linear or arbitrary phase response.

4.6 FINITE-BAND APPROXIMATIONS TO CONSTANT AMPLITUDE AND ARBITRARY PHASE RESPONSE

We shall solve this problem directly in terms of the arbitrary phase polynomials and recover the specific cases of those involving the equidistant linear phase and constant phase delay polynomials as specific cases.[3,4]

Consider an even-degree transfer function $S_{12}(p)$ of the form of equation (4.4.1) and of degree $2n$ which satisfies the constraints

$$S_{12}(\infty) = 0$$

$$|S_{12}(j\omega)| \leqslant 1 \tag{4.6.1}$$

$$|S_{12}(\pm j\omega_i)| = 1$$
$$\text{Arg } S_{12}(\pm j\omega_i) = \pm 2\psi(\omega_i) \qquad i = 1 \to n \tag{4.6.2}$$

We wish to obtain a closed-form solution to this problem and establish conditional requirements upon the ω_i and $\psi(\omega_i)$ such that $S_{12}(p)$ is a bounded real function.

From the amplitude condition it immediately follows that

$$|S_{12}(j\omega)|^2 = \frac{1}{1 + \left[\prod_{i=1}^{n}(\omega_i^2 - \omega^2)\Big/E(j\omega)\right]^2} \tag{4.6.3}$$

$$= \frac{E^2(j\omega)}{|G(j\omega)|^2} \tag{4.6.4}$$

where

$$G(j\omega) = -\prod_{i=1}^{n}(\omega_i^2 - \omega^2) + jE(j\omega) \qquad (4.6.5)$$

and

$$G(p) = D_n(p)D_n(-p) \qquad (4.6.6)$$

where $D_n(p)$ is a polynomial with left half-plane roots and possesses complex coefficients. Thus,

$$S_{12}(p) = \frac{E(p)}{D_n(p)D_n^*(p)} \qquad (4.6.7)$$

where $D_n^*(p)$ is $D_n(p)$ with the coefficients replaced by their complex conjugates.

From the phase conditions,

$$\text{Arg } D_n(j\omega_i) + \text{Arg } D_n^*(j\omega_i) = 2\psi(\omega_i) \qquad (4.6.8)$$

However, from equations (4.6.5) and (4.6.6),

$$\text{Arg } D_n(j\omega_i) - \text{Arg } D_n(-j\omega_i) = \frac{\pi}{2} \qquad (4.6.9)$$

or

$$\text{Arg } D_n(j\omega_i) - \text{Arg } D_n^*(j\omega_i) = \frac{\pi}{2} \qquad (4.6.10)$$

Combining (4.6.8) and (4.6.10) yields

$$\text{Arg } D_n(j\omega_i) = \psi(\omega_i) + \frac{\pi}{4} \qquad (4.6.11)$$

or

$$\text{Arg}[e^{-j\pi/4} D_n(j\omega_i)] = \psi(\omega_i) \qquad (4.6.12)$$

Consequently, using the results of Section 3.5,

$$e^{-j\pi/4} D_n(j\omega_i) = K_1 Q_n(j\omega_i) + jK_2 P_n(j\omega_i) \qquad (4.6.13)$$

or

$$D_n(p) = \frac{K_1(1+j)}{\sqrt{2}} Q_n(p) + \frac{K_2(1-j)}{\sqrt{2}} P_n(p) \qquad (4.6.14)$$

where $Q_n(p)$ and $P_n(p)$ are the arbitrary phase polynomials of the first and second kinds respectively and K_1 and K_2 are real constants. To determine these constants, equation (4.6.14) may be substituted into (4.6.5) and (4.6.6) to give

$$j[K_1^2 Q_n(p)Q_n(-p) - K_2^2 P_n(p)P_n(-p)]$$
$$+ K_1 K_2 [Q_n(p)P_n(-p) + Q_n(-p)P_n(p)]$$
$$= - \prod_{i=1}^{n}(\omega_i^2 + p^2) + jE(p) \quad (4.6.15)$$

Since the leading coefficients of $Q_n(p)$ and $P_n(p)$ are unity and $E(p)$ is restricted to be of maximum degree $2n - 2$, we have

$$K_1^2 = K_2^2$$
$$K_1 K_2 = \tfrac{1}{2} \quad (4.6.16)$$

giving

$$K_1 = K_2 = \frac{1}{\sqrt{2}} \quad (4.6.17)$$

Thus

$$2D_n(p) = (1+j)Q_n(p) + (1-j)P_n(p) \quad (4.6.18)$$

and

$$2E(p) = Q_n(p)Q_n(-p) - P_n(p)P_n(-p) \quad (4.6.19)$$

yielding the final result

$$S_{12}(p) = \frac{Q_n(p)Q_n(-p) - P_n(p)P_n(-p)}{Q_n^2(p) + P_n^2(p)} \quad (4.6.20)$$

From (4.6.3) and (4.6.7) it follows that for $S_{12}(p)$ to be a bounded real function $D_n(p)$ must be devoid of zeros in Re $p > 0$. To achieve these conditions we employ the technique used in Chapter 3.

It follows that $D_n(p)$ must satisfy a recurrence formula of the form

$$D_n(p) = (a_{n-1} + jpb_{n-1})D_{n-1}(p) + c_{n-1}(p^2 + \omega_{n-1}^2)D_{n-2}(p)$$
$$(4.6.21)$$

where a_{n-1}, b_{n-1} and c_{n-1} are real constants. Since the leading coefficients are unity we have,

$$c_{n-1} + jb_{n-1} = 1 \quad (4.6.22)$$

or

$$c_{n-1} = 1, \quad b_{n-1} = 0$$

and

$$a_{n-1} = \frac{D_n(p)}{D_{n-1}(p)} \bigg|_{p=j\omega_{n-1}}$$

$$= \frac{Q_n(p)}{Q_{n-1}(p)} \left[\frac{1 - jP_n(p)/Q_n(p)}{1 - jP_{n-1}(p)/Q_{n-1}(p)} \right] \bigg|_{p=j\omega_{n-1}}$$

$$= \alpha_{n-1} \frac{1 - [\beta_n/\omega_{n-1} + (\omega_n^2 - \omega_{n-1}^2)/\omega_{n-1}\alpha_{n-1}]}{1 - \beta_{n-1}/\omega_{n-1}} \quad (4.6.23)$$

which must be identically equal to

$$a_{n-1} = \alpha_{n-1} \frac{1 + [\beta_n/\omega_{n-1} + (\omega_n^2 - \omega_{n-1}^2)/\omega_{n-1}\alpha_{n-1}]}{1 + \beta_{n-1}/\omega_{n-1}} \quad (4.6.24)$$

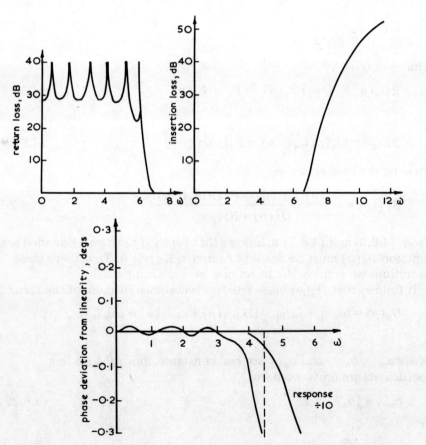

Figure 4.6.1 Computed response of transfer function interpolating to ideal amplitude and prescribed phase ($n = 12$)

giving the relationship between β_n, β_{n-1} and α_{n-1}. As before it may be shown that $P_n(p)$ belongs to a Hurwitz sequence if

$$a_r \geq 0 \qquad r = 1 \to n-1 \qquad (4.6.25)$$

i.e.

$$\alpha_r \geq 0, \qquad \left|\frac{\beta_r}{\omega_r}\right| \leq 1 \qquad r = 1 \to n-1 \qquad (4.6.26)$$

In Figure 4.6.1 an example of a 12th-degree transfer function is given where the phase interpolates to the characteristic

$$\psi(\omega) = \omega \left[1 + A \left(\frac{\omega}{\omega_c}\right)^N\right] \qquad (4.6.27)$$

and the ω_i themselves follow a similar distribution. From this return loss, insertion loss and phase characteristics the improvement in response over externally equalized Chebyshev filters can be appreciated for it may be shown that in addition to a 10th-degree Chebyshev filter to meet these amplitude requirements, an external 16th-degree reciprocal equalizer would be required to meet the phase response.

Before proceeding to consider the specific equidistant interpolation to a linear phase response and ideal constant amplitude characteristic, it is worthwhile to note that this approach is also capable of recovering the other more conventional classes of filters. For example, the Chebyshev response will be recovered if

$$\omega_i = \sin \frac{(2r-1)\pi}{2n}$$

and

$$\text{Arg } S_{12}(j\omega_i) = \tan^{-1} \frac{\eta}{\sin(\pi/n)} + 2 \sum_{q=2}^{i} \tan^{-1} \frac{\eta}{\sin(q\pi/n)} \qquad i = 1 \to n$$
$$(4.6.28)$$

using equation (4.2.17). A similar result follows for the elliptic function case.

For the equidistant interpolation to constant amplitude and linear phase response in this even-degree transfer function case, we use the equidistant constant phase delay polynomial $Q_n(p \mid \beta)$ as defined in Section 3.4 and its associated polynomial $P_n(p \mid \beta)$. Hence, the transfer function

$$S_{12}(p) = \frac{Q_n(p \mid \beta) Q_n(-p \mid \beta) - P_n(p \mid \beta) P_n(-p \mid \beta)}{Q_n^2(p \mid \beta) + P_n^2(p \mid \beta)} \qquad (4.6.29)$$

possesses the properties

$$S_{12}(j[2r-1]\beta) = e^{j(2r-1)\epsilon}, \qquad \beta = \tan \epsilon \qquad r = 1 \to n$$
$$S_{12}(\infty) = 0 \qquad (4.6.30)$$

For $S_{12}(p)$ to be bounded real it is necessary that condition (3.4.16) be satisfied which in the terminology used for the arbitrary phase polynomials ensures that

$$\alpha_r \geq 0 \qquad r = 1 \to n \qquad (4.6.31)$$

The additional condition is $|\beta_r/\omega_r| \leq 1$ which may be obtained by using equations (3.4.17) and (3.4.15) and expressing the result in the form of (3.5.22). Thus, including the particular normalization for the leading coefficients of $P_n(p\mid\beta)$ and $Q_n(p\mid\beta)$, we have

$$pP_n(p\mid\beta) = \frac{p^2 + (2n-1)^2\beta^2}{2n-1} Q_{n-1}(p\mid\beta) + \beta_n Q_n(p\mid\beta) \qquad (4.6.32)$$

where

$$\beta_n = -(2n-1)\beta^2$$

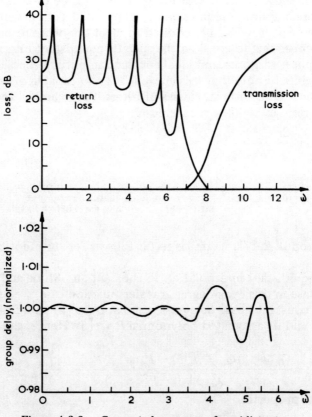

Figure 4.6.2 Computed response of equidistant interpolation to constant amplitude and phase delay ($n = 12$)

Hence the realizability condition is

$$|\beta| \leq 1 \qquad (4.6.33)$$

which ensures that $S_{12}(p)$ given in (4.6.29) is a bounded real function. A typical 12-degree response is shown in Figure 4.6.2.

For the odd-degree case let $S_{12}(p)$ be of degree $2n+1$. If $S_{12}(p)$ satisfies the same conditions as given in (4.6.1) and (4.6.2) with $S_{12}(0) = 1$, then $S_{12}(p)$ is uniquely defined apart from one parameter and using a similar method to that employed in the even-degree case, it may be shown that

$$S_{12}(p) = \frac{\mathrm{Ev}\{Q_n(-p)[(1+K)Q_{n+1}(p) - KQ_n(p)]\}}{Q_n(p)[(1+K)Q_{n+1}(p) - KQ_n(p)]} \qquad (4.6.34)$$

where K is an arbitrary constant. From the recurrence formula for $Q_n(p)$, $S_{12}(p)$ will be a bounded real function if

$$\alpha_r \geq 0 \qquad r = 1 \rightarrow n$$

and $\qquad (4.6.35)$

$$K \geq -1$$

Finally, the constant K may be chosen to provide either an extra amplitude or phase constraint similar to the maximally flat case described in Section 4.4.

4.7 PROTOTYPE SYNTHESIS PROCEDURE FOR TRANSMISSION TYPE FILTERS

The transfer functions described in the previous sections are bounded real functions (assuming that the realizability conditions are enforced) and consequently may be realized by resistively terminated lossless reciprocal two-ports using a cascade synthesis process. However, due to the fact that the numerator of the reflection coefficient is either an even or an odd polynomial it is always possible to realize the filter as a symmetrical device.

Using the bisection theorem, we may write

$$S_{12}(p) = \frac{Y_e - Y_o}{(1+Y_e)(1+Y_o)} \qquad (4.7.1)$$

where Y_e and Y_o are the even- and odd-mode admittances of a symmetrical network respectively. Y_e is the input admittance when an open-circuiting plane is inserted along the line of symmetry and Y_o the input admittance when a short-circuiting plane is inserted along the line of symmetry. For the odd-degree case, from equation (4.6.34) it

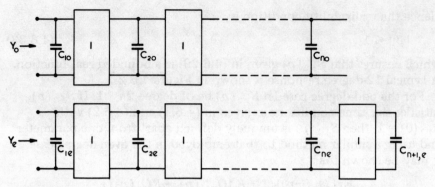

Figure 4.7.1 Realization of even- and odd-mode admittances (n odd)

immediately follows that

$$\frac{1+Y_o}{1-Y_o} = \pm \frac{Q_n(p)}{Q_n(-p)} \tag{4.7.2}$$

or

$$\frac{1}{Y_o} \text{ or } Y_o = \frac{\text{Od}[Q_n(p)]}{\text{Ev}[Q_n(p)]} \tag{4.7.3}$$

and

$$\frac{1}{Y_e} \text{ or } Y_e = \frac{\text{Od}[(1+K)Q_{n+1}(p) - KQ_n(p)]}{(1+K)Q_{n+1}(p) - KQ_n(p)} \tag{4.7.4}$$

The four combinations lead to transmission and reflection type filters and their duals. For one of the transmission type filters Y_o and Y_e should be chosen such that they both have transmission zeros at infinity. Since they are reactance functions, they may be synthesized as ladder networks as shown in Figure 4.7.1.

Combining the even- and odd-mode networks gives the complete lossless two-port operating between 1 Ω terminations with coupled capacitors as illustrated in Figure 4.7.2.

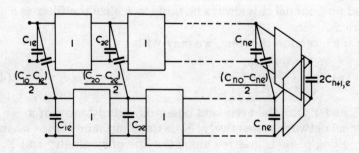

Figure 4.7.2 Two-port transmission filter for odd-degree networks

From this prototype network, devices using transformers may be constructed. If narrow-band band-pass devices are designed from this prototype, the transformers may be eliminated if

$$C_{ro} \geq C_{rc} \qquad r = 1 \to n \tag{4.7.5}$$

but resonant coupling between the two sides of the network will still be necessary and this is not particularly attractive from a physically realizability viewpoint. For this reason, the even-degree prototype assumes a more significant role.

Identifying (4.7.1) with (4.6.7) gives

$$Y_o = Y_e^* \tag{4.7.6}$$

and the zeros of $1 + Y_o$ are the zeros of $D_n(p)$. Writing

$$D_n(p) = \tfrac{1}{2}(1+j)Q_n(p) + \tfrac{1}{2}(1-j)P_n(p)$$
$$= E_1(p) + O_1(p) + j[E_2(p) + O_2(p)] \tag{4.7.7}$$

then

$$\begin{aligned}
E_1(p) &= \tfrac{1}{2}\text{Ev}[Q_n(p) + P_n(p)] \\
O_1(p) &= \tfrac{1}{2}\text{Od}[Q_n(p) + P_n(p)] \\
E_2(p) &= \tfrac{1}{2}\text{Ev}[Q_n(p) - P_n(p)] \\
O_2(p) &= \tfrac{1}{2}\text{Od}[Q_n(p) - P_n(p)]
\end{aligned} \tag{4.7.8}$$

Thus, if $Y_e(p)$ possesses a pole at infinity

$$Y_e(p) = \frac{E_1(p) + jO_2(p)}{O_1(p) + jE_2(p)} \qquad n \text{ even} \tag{4.7.9}$$

or

$$Y_e(p) = \frac{O_1(p) + jE_2(p)}{E_1(p) + jO_2(p)} \qquad n \text{ odd}$$

realizing $S_{12}(p)$ within a constant 90° phase shift. The synthesis cycle is commenced by extracting a shunt capacitor C_1 given by

$$\begin{aligned}
C_1 &= \left.\frac{E_1(p)}{pO_1(p)}\right|_{p=\infty} \qquad n \text{ even} \\
&= \left.\frac{O_1(p)}{pE_1(p)}\right|_{p=\infty} \qquad n \text{ odd}
\end{aligned} \tag{4.7.10}$$

and the remaining admittance $Y_1(p)$ is given by

$$Y_1(p) = Y(p) - C_1 p \tag{4.7.11}$$

Next, a frequency invariant reactance of susceptance jK_1 is extracted

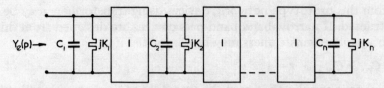

Figure 4.7.3 Realization of even-mode admittance for n even

to give a remainder

$$Y_2(p) = Y_1(p) - jK_1 \qquad (4.7.12)$$

and K_1 is chosen such that $Y_2(p)$ possess a zero at $p = \infty$. Therefore

$$K_1 = -jY_1(\infty) \qquad (4.7.13)$$

An inverter of unity impedance is extracted to give

$$Y_3(p) = \frac{1}{Y_2(p)} \qquad (4.7.14)$$

which possesses a pole at $p = \infty$. Thus the synthesis cycle may now be repeated until the remainder is zero. Hence, $Y_e(p)$ has been synthesized in the form shown in Figure 4.7.3.

The even-mode admittance only differs from the odd mode by changing $j \to -j$ and thereby changing the sign of the frequency invariant reactances. Thus, since the transfer matrix of a two-port in terms of its even- and odd-mode admittances is

$$\frac{1}{Y_e - Y_o} \begin{bmatrix} Y_e + Y_o & 2 \\ 2Y_e Y_o & Y_e + Y_o \end{bmatrix} \qquad (4.7.15)$$

for even- and odd-mode admittances $-jK_r$ and $+jK_r$ respectively, the

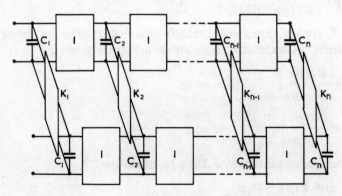

Figure 4.7.4 Two-port transmission filter for even-degree networks

transfer matrices for a typical coupling element becomes

$$\begin{bmatrix} 0 & j/K_r \\ jK_r & 0 \end{bmatrix} \qquad (4.7.16)$$

which represents an inverter of characteristic admittance K_r, and gives the realized lossless two-port operating between 1 Ω terminates as shown in Figure 4.7.4.

Unfortunately, there do not appear to be any explicit formulas for element values in this network for any particular case of transfer function apart from the degenerate Chebyshev and maximally flat amplitude cases. However, the synthesis process is very simple as illustrated in the following example.

Consider the transfer function (4.6.29) for the 4th-degree case [3.3] where

$$S_{12}(p) = \frac{\text{Im}[D_2(p)D_2(-p)]}{D_2(p)D_2(p)} \qquad (4.7.17)$$

and

$$D_2(p) = (1+j)Q_2(p/\beta) + (1-j)P_2(p/\beta)$$
$$= 1 + \beta^4 + \tfrac{4}{3}(1-\beta^2)p + \tfrac{2}{3}p^2 + j[1-\beta^4 + \tfrac{2}{3}(1+\beta^2)p] \qquad (4.7.18)$$

since

$$Q_2(p\mid\beta) = 1 + \left(1 - \frac{\beta^2}{3}\right)p + \frac{p^2}{3}$$

and $\qquad (4.7.19)$

$$P_2(p\mid\beta) = \beta^4 + [\tfrac{1}{3} - \beta^2]p + \frac{p^2}{3}$$

The even-mode admittance $Y_e(p)$ is therefore

$$Y_e(p) = \frac{1 + \beta^4 + \tfrac{2}{3}p^2 + jp\tfrac{2}{3}(1+\beta^2)}{(1-\beta^2)[\tfrac{2}{3}p + j(1+\beta^2)]} \qquad (4.7.20)$$

and

$$C_1 = \left.\frac{Y_e(p)}{p}\right|_{p=\infty} = \frac{1}{2(1-\beta^2)} \qquad (4.7.21)$$

leaving

$$Y_1(p) = Y_e(p) - C_1 p$$
$$= \frac{2(1+\beta^4) + jp\tfrac{1}{3}(1+\beta^2)}{(1-\beta^2)[\tfrac{4}{3}p + j(1+\beta^2)]} \qquad (4.7.22)$$

Removal of the constant imaginary part of infinity gives

$$K_1 = -jY_1(\infty) = \frac{1+\beta^2}{8(1-\beta^2)} \qquad (4.7.23)$$

and

$$Y_2(p) = Y_1(p) - jK_1$$

$$= \frac{9 + 2\beta^2 + 9\beta^4}{8(1-\beta^2)[\tfrac{4}{3}p + j(1+\beta^2)]} \qquad (4.7.24)$$

Extracting an inverter of unity characteristic admittance gives

$$Y_2(p) = C_2 p + jK_2 \qquad (4.7.25)$$

with

$$C_2 = \frac{32(1-\beta^2)}{3(9 + 2\beta^2 + 9\beta^4)}$$

and

$$K_2 = \frac{8(1-\beta^4)}{9 + 2\beta^2 + 9\beta^4} \qquad (4.7.26)$$

If $|\beta| \leqslant 1$ then the capacitors are non-negative. In fact for any bounded real transfer function realized in this manner the capacitors must be non-negative. However, the characteristic admittances of the coupling inverters K_r may in general be positive or negative. In this particular example we see that the realizability condition $|\beta| \leqslant 1$ also ensures that $K_1, K_2 \geqslant 0$. For this particular type of transfer function, for low-degree

Table 4.7.1
Element values for maximally flat amplitude and phase transmission filter

N	4	6	8	10	12	14
C_1	0.5000	0.2222	0.1250	0.0800	0.0556	0.0408
K_1	0.1250	0.0370	0.0156	0.0080	0.0046	0.0029
C_2	1.1852	0.5956	0.3519	0.2305	0.1621	0.1200
K_2	0.8889	0.2248	0.0794	0.0356	0.0188	0.0111
C_3		0.8281	0.5142	0.3527	0.2547	0.1916
K_3		0.7916	0.2868	0.1165	0.0546	0.0290
C_4			0.6583	0.4365	0.3253	0.2508
K_4			0.7187	0.3259	0.1514	0.0750
C_5				0.5652	0.3778	0.2945
K_5				0.6605	0.3470	0.1823
C_6					0.5083	0.3352
K_6					0.6112	0.3546
C_7						0.4696
K_7						0.5677

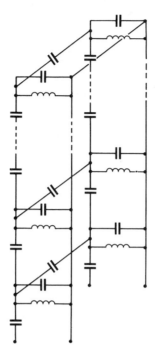

Figure 4.7.5 Band-pass transmission filter constructed from low-pass prototype

cases ($N \leq 8$), by direct synthesis it may be shown that for $|\beta| \leq 1$, $K_r \geq 0$ and it appears that this may also be true for arbitrary degree. In the degenerate maximally flat case, as described in Section 4.4, $K_r \geq 0$ for all r as can be seen from the table of element values given below (Table 4.7.1).

The significance of the coupling elements being non-negative may be appreciated if the prototype is used for the construction of narrow-band band-pass filters. Using the capacitively coupled resonant cavity technique the network shown in Figure 4.7.5 may be obtained as an approximate band-pass version of the prototype. For realizability, all elements must be non-negative which implies that $K_r \geq 0$ for all r.

For high-degree networks, loss of accuracy could occur during the synthesis process due to the loss of significant figures at each cycle. However, for the maximally flat case, even for networks of degree up to 100, it has been found that no loss of significant figures occurs. This is obviously due to the sign of the frequency invariant reactances which are extracted and inherently indicates that this type of network realization for these particular classes of transfer functions is not very sensitive to changes in element values.

CHAPTER 5

Amplitude Approximations for Distributed Networks

5.1 INTRODUCTION

For a finite commensurated distributed element filter we have (ω normalized)

$$|S_{12}(j\tan\omega)|^2 = \frac{\sum_{1}^{m} a_i \tan^2\omega}{\sum_{1}^{n} b_i \tan^2\omega} \tag{5.1.1}$$

and inherently all response characteristics must be periodic in ω. Again we shall be concerned with the quasi low-pass and later quasi high-pass or band-pass response characteristics. In the distributed domain, an ideal low-pass amplitude characteristic will be as shown in Figure 5.5.1.

However, in terms of the variable $\tan\omega$ as compared to ω, there is no difference between the solution to the amplitude problem in the distributed and lumped domains. Thus the maximally flat, Chebyshev, inverse Chebyshev and elliptic function characteristics for distributed filters are identical to the results obtained in Chapter 2 apart from the application of the complex frequency transformation

$$p \to \frac{\tanh p}{\alpha} = \frac{t}{\alpha} \tag{5.1.2}$$

with

$$\alpha = \tan\omega_0 \tag{5.1.3}$$

which maps the point $\omega = 1$ into the point $\omega = \omega_0$ in the distributed domain. Thus, in theory, there is no apparent reason for pursuing the distributed case any further. However, if one looks at the physical realization of distributed filters, then certain problems arise.

From the lumped element prototype ladder networks, application of the frequency transformation (5.1.2) leads to the conclusion*
that capacitors may be replaced by open-circuited lines and

---

*The short- and open-circuited input impedances of a transmission line are Zt and Z/t respectively.

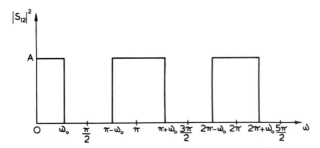

Figure 5.1.1 Ideal low-pass distributed amplitude characteristic

inductors may be replaced by short-circuited lines; all these lines or stubs being of the same electrical length. However, it is impossible to construct such a structure since all of the shunt and series connections are required to be connected at the same physical point. It is necessary to separate these stubs with lengths of transmission lines known as unit elements (u.e.s) described by transfer matrices of the form

$$\frac{1}{\sqrt{1-t^2}} \begin{bmatrix} 1 & Zt \\ Yt & 1 \end{bmatrix} \tag{5.1.4}$$

where Z is the characteristic impedance. Using Kuroda transformations, e.g.[5.1], these u.e.s may be introduced to be between every pair of stubs without changing the amplitude response. Although redundant in an electrical or mathematical sense, these u.e.s are necessary in a physical sense. Accepting this fact, one must ask the question whether or not these necessary additional unit elements may be employed to improve the selectivity of the filter and immediately one can conclude that this must be the case since they can increase the degree of the transfer function. This immediately poses the question as to whether a cascade of u.e.s alone can produce a low-pass filter characteristic. Again the answer is in the affirmitive and considerable attention will be paid to this basic distributed prototype filter in this chapter, eventually leading to explicit design formulas for Chebyshev filters.

After briefly considering the extension of the results on the basic distributed prototype filter to the mixed stub—u.e. filter, the particular band-pass response associated with the interdigital filter will be investigated. Again this is justified due to the physical importance of the realization in the form of a single n-wire coupled line and explicit design formulas are obtained.

In the final part of the chapter a novel method of approximation is introduced enabling cascades of u.e.s to approximate arbitrarily prescribed amplitude characteristics and an algorithm is developed to allow the characteristic impedances to be obtained without recourse to formal synthesis techniques.

5.2 STEPPED IMPEDANCE TRANSMISSION LINE FILTERS WITH MAXIMALLY FLAT AND CHEBYSHEV RESPONSE CHARACTERISTICS

For a cascade of n unit elements (u.e.s) as shown in Figure 5.2.1, by analysis it may be shown that

$$S_{12}(t) = \frac{(1-t^2)^{\frac{1}{2}n}}{D_n(t)} \qquad (5.2.1)$$

where $D_n(t)$ is a Hurwitz polynomial in $t = \tanh p$ and $|S_{12}(j \tan \omega)|^2 \leq 1$. Thus,

$$|S_{12}(j \tan \omega)|^2 = \frac{1}{1 + P_n(\sin^2 \omega)} \qquad (5.2.2)$$

where $P_n(x)$ is a polynomial and $P_n(x) \geq 0$ for $0 \leq x \leq \infty$.

For a maximally flat response about the origin with unity gain it is apparent that $|S_{12}(j \tan \omega)|^2$ must be of the form

$$|S_{12}(j \tan \omega)|^2 = \frac{1}{1 + (\sin \omega/\alpha)^{2n}} \qquad (5.2.3)$$

and if $\alpha = \sin \omega_0$, $\omega = \omega_0$ results in the 3 dB point. This constrains the first $(2n-1)$ derivatives to vanish at the origin but only the first derivative vanishes at $\omega = \frac{1}{2}\pi$.

For an equiripple or Chebyshev response, the transfer function will be of the form

$$|S_{12}(j \tan \omega)|^2 = \frac{1}{1 + \epsilon^2 T_n^2(\sin \omega/\alpha)} \qquad (5.2.4)$$

where $T_n(x) = \cos(n \cos^{-1} x)$, resulting in the response shown in Figure 5.2.2.

This response in optimally equiripple in the passband and the first derivative vanishes at $\omega = \frac{1}{2}\pi$. The question must be considered as to whether, within the restriction that $1/|S_{12}|^2$ is polynomial in $\sin^2 \omega$, this response is optimum in the sense of minimizing the loss in the passband whilst maintaining a level greater than a prescribed value in the stopband. Using techniques similar to those employed in Chapter 1,

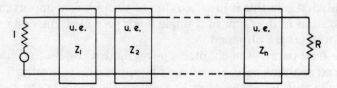

Figure 5.2.1   Unit element filter

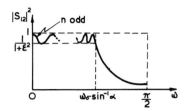

Figure 5.2.2 Chebyshev response for distributed prototype filter

it may be shown that this is indeed optimum in the sense described. However, this result should be contrasted with the case considered in the next chapter when $|S_{12}|^2$ is itself a polynomial in $\sin^2 \omega$.

To obtain the transfer function $S_{12}(t)$ and the corresponding $S_{11}(t)$, the Hurwitz factorization of (5.2.3) and (5.2.4) must be obtained. From (5.2.3) we have

$$|S_{12}|^2 = \frac{1}{1 + (\sin \omega/\alpha)^{2n}} \tag{5.2.5}$$

and

$$|S_{11}|^2 = \frac{\sin^{2n} \omega}{\alpha^{2n} + \sin^{2n} \omega} \tag{5.2.6}$$

The poles occur when

$$\sin^{2n} \omega = -\alpha^{2n}$$

or

$$\sin \omega = \alpha e^{j\theta_r} \qquad \theta_r = \frac{(2r-1)\pi}{2n} \tag{5.2.7}$$

Alternatively, these occur when

$$e^{j\omega} = \cos \omega + j \sin \omega = \sqrt{1 - \alpha^2 e^{j2\theta_r}} + j\alpha e^{j\theta_r} \tag{5.2.8}$$

Defining a new variable

$$z = \frac{1-t}{1+t} = e^{-2p}$$

and using (5.2.6) we have

$$S_{11}(z) = \frac{(1-z)^n}{\prod_{r=1}^{n} [(\sqrt{1 - \alpha^2 e^{j2\theta_r}} - j\alpha e^{j\theta_r}) - z(\sqrt{1 - \alpha^2 e^{j2\theta_r}} + j\alpha e^{j\theta_r})]} \tag{5.2.9}$$

which satisfies the bounded real condition

$$|S_{11}(z)| \leqslant 1 \qquad |z| \leqslant 1 \tag{5.2.10}$$

if $\alpha \leqslant 1$, since all of the zeros are at $z = 1$ and

$$S_{11}(-1) = \frac{1}{\sqrt{1+\alpha^{2n}}} \tag{5.2.11}$$

For $S_{12}(z)$, half the zeros are at $z = 0$ and the other half at $z = \infty$, hence

$$S_{12}(z) = \frac{(2\alpha)^n z^{\frac{1}{2}n}}{\prod_{r=1}^{n}[(\sqrt{1-\alpha^2}\,e^{j2\theta_r} - j\alpha e^{j\theta_r}) - z(\sqrt{1-\alpha^2}\,e^{j2\theta_r} + j\alpha e^{j\theta_r})]} \tag{5.2.12}$$

For the Chebyshev filter

$$|S_{12}|^2 = \frac{1}{1+\epsilon^2 T_n^2(\sin\omega/\alpha)} \tag{5.2.13}$$

and

$$|S_{11}|^2 = \frac{\epsilon^2 T_n^2(\sin\omega/\alpha)}{1+\epsilon^2 T_n^2(\sin\omega/\alpha)} \tag{5.2.14}$$

The poles occur when

$$\cos^2\left(n\cos^{-1}\frac{\sin\omega}{\alpha}\right) = -\frac{1}{\epsilon^2} \tag{5.2.15}$$

or

$$\sin\omega = \alpha\cos(\sin^{-1}j\eta + \theta_r) \tag{5.2.16}$$

where

$$\eta = \sinh\left(\frac{1}{n}\sinh^{-1}\frac{1}{\epsilon}\right) \tag{5.2.17}$$

and

$$\theta_r = \frac{(2r-1)\pi}{2n}$$

Since the zeros of $S_{11}(z)$ occur at

$$\sin\omega = \alpha\cos\theta_r \tag{5.2.18}$$

we have the bounded real reflection coefficient

$$S_{11}(z) = \prod_{r=1}^{n} \frac{(\sqrt{1-\alpha^2\cos^2\theta_r} - j\alpha\cos\theta_r) - z(\sqrt{1-\alpha^2\cos^2\theta_r} + j\alpha\cos\theta_r)}{(\sqrt{1-\alpha^2\cos^2\psi_r} - j\alpha\cos\psi_r) - z(\sqrt{1-\alpha^2\cos^2\psi_r} + j\alpha\cos\psi_r)} \tag{5.2.19}$$

where

$$\psi_r = \sin^{-1} j\eta + \theta_r \qquad \theta_r = \frac{(2r-1)\pi}{2n} \qquad (5.2.20)$$

Similarly,

$$S_{12}(z) = \frac{(2\alpha)^n z^{\frac{1}{2}n} \prod_{r=1}^{n} [\eta^2 + \sin^2(r\pi/n)]^{1/2}}{\prod_{r=1}^{n} [(\sqrt{1-\alpha^2} \cos^2 \psi_r - j\alpha \cos \psi_r) - z(\sqrt{1-\alpha^2} \cos^2 \psi_r + j\alpha \cos \psi_r)]}$$

$$(5.2.21)$$

5.3 EXPLICIT FORMULAS FOR ELEMENT VALUES IN CHEBYSHEV STEPPED IMPEDANCE TRANSMISSION LINE FILTERS

We shall now concentrate upon the Chebyshev response characteristic and recover the maximally flat response as a degenerate case. The material in this section is an extension of the ideas used in Chapter 2 for the derivation of explicit formulas for element values in Chebyshev ladder networks. However, from equations (5.2.20) and (5.2.21) it may be noticed that the network functions are dependent upon the bandwidth scaling factor $\alpha = \sin \omega_0$ which does not occur as a simple frequency scaling factor as in the lumped case. Furthermore, these network functions are irrational functions of α and no transformation of variable can make them rational in any other auxiliary variable. Thus, it appears unlikely that any exact solution for explicit formulas for the element values can be achieved. However, $\alpha = \sin \omega_0$ is less than unity and for most practical values of bandwidth where a reasonable attenuation is required at $\omega = \frac{1}{2}\pi$, ω_0 will be less than $\frac{1}{4}\pi$. As α becomes very small, the response characteristic approaches the lumped case for which we have already obtained explicit formulas. Thus it appears that an obvious approach is to try to obtain explicit formulas for element values as a power series expansion in α and the first few terms should suffice for any design of moderate bandwidth.

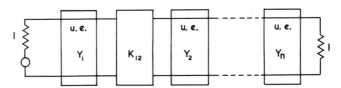

Figure 5.3.1 Basic distributed prototype incorporating impedance inverters

In order to follow the line of development used in Chapter 2, we first make the trivial modification to the prototype of introducing impedance inverters between every u.e. Once the explicit formulas have been obtained it is a simple task to transform these out of the network. This modification is illustrated in Figure 5.3.1 and also results in $S_{12}(z)$ being modified to

$$S_{12}(z) = -\frac{(2j\alpha)^n z^{1/2 n} \prod_{r=1}^{n} [\eta^2 + \sin^2(r\pi/n)]^{1/2}}{\prod_{r=1}^{n} [(\sqrt{1-\alpha^2 \cos^2 \psi_r} - j\alpha \cos \psi_r) - z(\sqrt{1-\alpha^2 \cos^2 \psi_r} + j\alpha \cos \psi_r)]} \quad (5.3.1)$$

Evaluation at the specific frequency given by

$$\sin \omega = -\alpha \cos \frac{\pi}{2n} \quad (5.3.2)$$

yields

$$|S_{12}|^2 = 1 \quad (5.3.3)$$

and

$$T_{12} = S_{12}(z)|_{\sin \omega = -\alpha \cos(\pi/2n)}$$

$$= \frac{\prod_{r=1}^{n} [\eta^2 + \sin^2(r\pi/n)]^{1/2}}{\prod_{r=1}^{n} [\cos \psi_r \sqrt{1-\alpha^2 \cos^2(\pi/2n)} + \cos(\pi/2n)\sqrt{1-\alpha^2 \cos^2 \psi_r}]} \quad (5.3.4)$$

Since the poles and zeros are independent of α, we may therefore write T_{12} in the form

$$T_{12} = \left[\frac{H_n(-\lambda)}{H_n(\lambda)}\right]^{1/2} e^{-Q(\alpha, \lambda)} \quad (5.3.5)$$

where $\lambda = +j\eta$ and the zeros of $H_n(\lambda)$ occur when

$$\cos\left[\sin^{-1}(+\lambda) + \frac{(2r-1)\pi}{2n}\right] = -\cos \frac{\pi}{2n} \quad (5.3.6)$$

or

$$\lambda = -\sin \frac{(r-1)\pi}{n} \quad (5.3.7)$$

Hence,

$$H_n(\lambda) = \prod_{r=1}^{n-1} \left(\lambda + \sin \frac{r\pi}{n}\right) \tag{5.3.8}$$

is a strict Hurwitz polynomial in λ.

Also, from (5.3.4) and (5.3.5)

$$Q(\alpha,\lambda) = \sum_{r=1}^{n} \ln\left[\frac{\cos\psi_r\sqrt{1-\alpha^2\cos^2(\pi/2n)} + \cos(\pi/2n)\sqrt{1-\alpha^2\cos^2\psi_r}}{\cos\psi_r + \cos(\pi/2n)}\right] \tag{5.3.9}$$

$$= +\frac{\alpha^2}{2}\cos\frac{\pi}{2n}\sum_{r=1}^{n}\cos\psi_r + O(\alpha^4) \qquad |\alpha|<1 \tag{5.3.10}$$

where $O(\alpha^4)$ contains terms of α^4 and above.

Now using equation (4.2.9),

$$\sum_{r=1}^{n}\cos\psi_r = \sum_{r=1}^{n}\sqrt{1+\eta^2}\cos\frac{(2r-1)\pi}{2n} - j\eta\sin\frac{(2r-1)\pi}{2n}$$

$$= -\frac{j\eta}{\sin(\pi/2n)} \tag{5.3.11}$$

therefore

$$Q(\alpha,\lambda) = +\frac{\lambda\alpha^2\cot(\pi/2n)}{2} + O(\alpha^4) \tag{5.3.12}$$

Furthermore, T_{12} is a bounded real unitary function in λ since it is analytic in Re $\lambda \geq 0$ including $\lambda = \infty$ approached from Re $\lambda > 0$, and must therefore be realizable as a cascade of passive all-pass sections in λ, each with a bounded unitary transfer function S_{12}^r given by

$$S_{12}^r = \left[\frac{\sin(r\pi/n) - \lambda}{\sin(r\pi/n) + \lambda}\right]^{1/2} e^{-[+\beta_r\lambda\alpha^2 + O_r(\alpha^4)]} \tag{5.3.13}$$

where

$$\beta_r \geq 0 \qquad r = 1 \to n-1 \tag{5.3.14}$$

and

$$\sum_{r=0}^{n}\beta_r = +\tfrac{1}{2}\cot\frac{\pi}{2n} \tag{5.3.15}$$

Turning now to the reflection coefficient we specifically recover the dependence upon η as

$$S_{11}(z,\eta) = \prod_{r=1}^{n} \frac{\begin{array}{c}(\sqrt{1-\alpha^2\cos^2\theta_r} - j\alpha\cos\theta_r) \\ -z(\sqrt{1-\alpha^2\cos^2\theta_r} + j\alpha\cos\theta_r)\end{array}}{\begin{array}{c}(\sqrt{1-\alpha^2\cos^2\psi_r} - j\alpha\cos\psi_r) \\ -z(\sqrt{1-\alpha^2\cos^2\psi_r} + j\alpha\cos\psi_r)\end{array}}$$

(5.3.16)

Choosing a value of η such that $S_{12}(z) = 0$, i.e.

$$\eta = j\sin\frac{q\pi}{n} \qquad (5.3.17)$$

we have

$$|S_{11}|^2 = 1 \qquad (5.3.18)$$

and in particular it follows from (5.3.16) that

$$S_{11}\left(z, j\sin\frac{q\pi}{n}\right) = \prod_{r=1}^{q} \left[\frac{\begin{array}{c}(\sqrt{1-\alpha^2\cos^2\theta_r} + j\alpha\cos\theta_r) \\ -z(\sqrt{1-\alpha^2\cos^2\theta_r} - j\alpha\cos\theta_r)\end{array}}{\begin{array}{c}(\sqrt{1-\alpha^2\cos^2\theta_r} - j\alpha\cos\theta_r) \\ -z(\sqrt{1-\alpha^2\cos^2\theta_r} + j\alpha\cos\theta_r)\end{array}}\right]$$

with

$$\theta_r = \frac{(2r-1)\pi}{2n} \qquad (5.3.19)$$

is a function of degree q in n. A similar result holds if $\eta = -j\sin(q\pi/n)$. Consequently we may choose the value of the characteristic admittances of the inverters separating the u.e.s to possess a value

$$K_{r,r+1} = \sqrt{\eta^2 + \sin^2\frac{r\pi}{n}} M(\eta,\alpha) \qquad (5.3.20)$$

where $M(\eta,\lambda)$ is finite at $\eta = \pm j\sin(r\pi/n)$, since setting the qth inverter to zero admittance produces a unitary reflection coefficient of degree q in z.

Thus, we have been able to isolate the position in the network at which the transmission zeros in the auxiliary parameter η occur.

Applying this to the result described by equation (5.3.13) implies that at the frequency $\sin\omega = -\alpha\cos(\pi/2n)$, the network must reduce to a cascade of two-ports described by the overall transfer matrix

$$\prod_{r=0}^{n} \begin{bmatrix} \cos\gamma_r & j\sin\gamma_r \\ j\sin\gamma_r & \cos\gamma_r \end{bmatrix} \qquad (5.3.21)$$

with

$$\gamma_r = \cot^{-1}\frac{\sin(r\pi/n)}{\eta} + \beta_r\eta\alpha^2 + O'_r(\alpha t) \tag{5.3.22}$$

Each two-port described by (5.3.21) may be decomposed by extracting shunt and then series elements from either end to leave a remaining section described by

$$\begin{bmatrix} 1 & -jX'_r \\ 0 & 1 \end{bmatrix}\begin{bmatrix} 1 & 0 \\ -jB'_r & 1 \end{bmatrix}\begin{bmatrix} \cos\gamma_r & j\sin\gamma_r \\ j\sin\gamma_r & \cos\gamma_r \end{bmatrix}\begin{bmatrix} 1 & 0 \\ -jB_{r+1} & 1 \end{bmatrix}$$

$$\begin{bmatrix} 1 & -jX_{r+1} \\ 0 & 1 \end{bmatrix} = \begin{bmatrix} A & jB \\ jC & D \end{bmatrix} \tag{5.3.23}$$

with

$$AD - BC = 1 \tag{5.3.24}$$

$$C = -(B'_r + B_{r+1})\cos\gamma_r + (1 - B'_r B_{r+1})\sin\gamma_r \tag{5.3.25}$$

$$A = \cos\gamma_r + B'_r \sin\gamma_r + X_{r+1} C \tag{5.3.26}$$

$$D = \cos\gamma_r + B_{r+1} \sin\gamma_r + X'_r C \tag{5.3.27}$$

which we may identify as an impedance inverter with characteristic admittance

$$K_{r,\,r+1} = -C \tag{5.3.28}$$

and

$$A = D = 0 \tag{5.3.29}$$

Consequently the u.e.s separating the impedance inverters possessing a transfer matrix at $\sin\omega = -\alpha\cos(\pi/2n)$ of

$$\begin{bmatrix} \sqrt{1-\alpha^2\cos^2\dfrac{\pi}{2n}} & -jZ_r\alpha\cos\dfrac{\pi}{2n} \\ -jY_r\alpha\cos\dfrac{\pi}{2n} & \sqrt{1-\alpha^2\cos^2\dfrac{\pi}{2n}} \end{bmatrix} \tag{5.3.30}$$

must also be described by the transfer matrix

$$\begin{bmatrix} 1 & jX_r \\ 0 & 1 \end{bmatrix}\begin{bmatrix} 1 & 0 \\ j(B_r + B'_r) & 1 \end{bmatrix}\begin{bmatrix} 1 & jX'_r \\ 0 & 1 \end{bmatrix}$$

$$= \begin{bmatrix} 1 - X_r(B_r + B'_r) & j[(X_r + X'_r) - X_r X'_r(B_r + B'_r)] \\ j(B_r + B'_r) & 1 - X'_r(B_r + B'_r) \end{bmatrix}$$

$$\tag{5.3.31}$$

yielding the relationships

$$X'_r = X_r \tag{5.3.32}$$

$$1 - X_r(B_r + B'_r) = \sqrt{1 - \alpha^2 \cos^2 \frac{\pi}{2n}} \tag{5.3.33}$$

and

$$Y_r = \frac{-(B_r + B'_r)}{\alpha \cos(\pi/2n)} \tag{5.3.34}$$

From equations (5.3.26)–(5.3.29) we have

$$B'_r = -\cot \gamma_r + \frac{X_{r+1} K_{r,r+1}}{\sin \gamma_r} \tag{5.3.35}$$

$$B_r = -\cot \gamma_{r-1} + \frac{X_{r-1} K_{r-1,r}}{\sin \gamma_{r-1}} \tag{5.3.36}$$

and substituting into equation (5.3.34)

$$Y_r \alpha \cos \frac{\pi}{2n} = \cot \gamma_r + \cot \gamma_{r-1} - \frac{X_{r+1} K_{r,r+1}}{\sin \gamma_r} - \frac{X_{r-1} K_{r-1,r}}{\sin \gamma_{r-1}} \tag{5.3.37}$$

Using equations (5.3.25)–(5.3.29),

$$K_{r,r+1} = \frac{1 - X_r X_{r+1} K^2_{r,r+1}}{\sin \gamma_r} \tag{5.3.38}$$

and from (5.3.33) and (5.3.34)

$$X_r = \frac{\sqrt{1 - \alpha^2 \cos^2(\pi/2n)} - 1}{Y_r \alpha \cos(\pi/2n)} \tag{5.3.39}$$

Using the latter three equations, it immediately follows that

$$Y_r = \frac{\sin(r\pi/n) + \sin[(r-1)\pi/n]}{\alpha \eta \cos(\pi/2n)} + O_{1r}(\alpha)$$

$$= \frac{2 \sin[(2r-1)\pi/2n]}{\alpha \eta} + O_{1r}(\alpha) \tag{5.3.40}$$

$$X_r = \frac{\eta \alpha^2 \cos(\pi/2n)}{4 \sin[(2r-1)\pi/2n]} + O_{2r}(\alpha^4) \tag{5.3.41}$$

$$K_{r,r+1} = \frac{1}{\sin \gamma_r} + O_{3r}(\alpha^4) \tag{5.3.42}$$

In making the above network decomposition at the frequency

$\sin\omega = -\alpha\cos(\pi/2n)$, care must be taken to ensure that the initial and final conditions at the input and output of the network are correct. These are

$$B'_0 = B_{n+1} = 0 \tag{5.3.43}$$

and therefore from (5.3.26) and (5.3.42)

$$\frac{X_1}{\cos\gamma_0 \sin\gamma_0} = O_4(\alpha^4) \tag{5.3.44}$$

and using (5.3.41)

$$\gamma_0 = \frac{\pi}{2} + \frac{\eta\alpha^2 \cot(\pi/2n)}{4} + O_5(\alpha^4) \tag{5.3.45}$$

Similarly

$$\gamma_n = \gamma_0$$

and hence

$$\beta_0 = \beta_n = \frac{+\cot(\pi/2n)}{4} \tag{5.3.46}$$

Consequently, from (5.3.14) and (5.3.15)

$$\beta_r = 0 \qquad r = 1 \to n-1 \tag{5.3.47}$$

and

$$\gamma_r = \cot^{-1}\frac{\sin(r\pi/n)}{\eta} + O_{1r}(\alpha^4) \tag{5.3.48}$$

From (5.3.47) and (5.3.42),

$$K_{r,r+1} = \frac{\sqrt{\eta^2 + \sin^2(r\pi/n)}}{\eta} + O_{3r}(\alpha^4) \qquad r = 1 \to n-1 \tag{5.3.49}$$

Scaling Y_1 and Y_n by K_{01} and $K_{n,n+1}$ respectively, substitution into (5.3.37) yields

$$Y_r = \frac{2\sin[(2r-1)\pi/2n]}{\eta\alpha} - \frac{\alpha}{4\eta}\left\{\frac{\eta^2 + \sin^2(r\pi/n)}{\sin[(2r+1)\pi/2n]}\right.$$

$$\left. + \frac{\eta^2 + \sin^2[(r-1)\pi/n]}{\sin[(2r-3)\pi/2n]}\right\} + O(\alpha^3) \qquad r = 1 \to n \tag{5.3.50}$$

which establishes the design equations for the Chebyshev response.

The design equations for the maximally flat case may be recovered by applying the transformation $\eta \to \beta/\alpha$ then by allowing $\alpha \to 0$, and finally replacing β by α. Hence,

$$K_{r,r+1} = 1 \tag{5.3.51}$$

$$Y_r = \frac{2\sin[(2r-1)\pi/2n]}{\alpha} - \frac{\alpha}{4}\left\{\frac{1}{\sin[(2r+1)\pi/2n]} + \frac{1}{\sin[(2r-3)\pi/2n]}\right\} + O(\alpha^3)$$

$$= \frac{2\sin[(2r-1)\pi/2n]}{\alpha}\left\{1 - \frac{\cos(\pi/n)\alpha^2}{4\sin[(2r-3)\pi/2n]\sin[(2r+1)\pi/2n]}\right\} + O(\alpha^3) \tag{5.3.52}$$

5.4 SUMMARY OF RESULTS FOR CHEBYSHEV AND MAXIMALLY FLAT DISTRIBUTED PROTOTYPE FILTERS

With regard to the stepped impedance distributed prototype filter consisting of n u.e.s separated by impedance inverters of characteristic admittance $K_{r,r+1}$ as shown in Figure 5.4.1, we may summarize the design equations derived in the previous section.

A. Maximally Flat Quasi Low Pass

$$|S_{12}|^2 = \frac{1}{1 + (\sin\omega/\alpha)^{2n}} \tag{5.4.1}$$

For $r = 1 \to n$

$$Y_r = \frac{2\sin[(2r-1)\pi/2n]}{\alpha}\left\{1 - \frac{\cos(\pi/n)\alpha^2}{4\sin[(2r-3)\pi/2n]\sin[(2r+1)\pi/2n]}\right\} + O(\alpha^3) \tag{5.4.2}$$

$$K_{r,r+1} = 1 \tag{5.4.3}$$

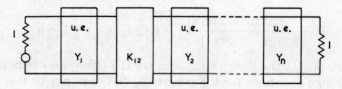

Figure 5.4.1 Basic distributed prototype

B. Equiripple Quasi Low Pass

$$|S_{12}|^2 = \frac{1}{1 + \epsilon^2 T_n^2(\sin\omega/\alpha)} \qquad (5.4.4)$$

For $r = 1 \to n$

$$Y_r = \frac{2\sin[(2r-1)\pi/2n]}{\eta\alpha} - \frac{\alpha}{4\eta}\left\{\frac{\eta^2 + \sin^2(r\pi/n)}{\sin[(2r+1)\pi/2n]}\right.$$

$$\left. + \frac{\eta^2 + \sin^2[(r-1)\pi/n]}{\sin[(2r-3)\pi/2n]}\right\} + O(\alpha^3) \qquad (5.4.5)$$

For $r = 1 \to n-1$

$$K_{r,r+1} = \frac{\sqrt{\eta^2 + \sin^2(r\pi/n)}}{\eta} + O'(\alpha^3) \qquad (5.4.6)$$

where

$$\eta = \sinh\left(\frac{1}{n}\sinh^{-1}\frac{1}{\epsilon}\right) \qquad (5.4.7)$$

The inverters may now be transformed out of the network to give a pure cascade of u.e.s as shown in Figure 5.4.2. The resulting values for the characteristic admittances are:

C. Maximally Flat without Inverters ($R_L = 1$)

$$\begin{array}{ll} Y_r = g_r & r \text{ odd} \\ 1/g_r & r \text{ even} \end{array} \qquad (5.4.8)$$

For $r = 1 \to n$

$$g_r = \frac{2\sin[(2r-1)\pi/2n]}{\alpha}$$

$$\left\{1 - \frac{\cos(\pi/n)\alpha^2}{4\sin[(2r-3)\pi/2n]\sin[(2r+1)\pi/2n]}\right\} + O(\alpha^3) \qquad (5.4.9)$$

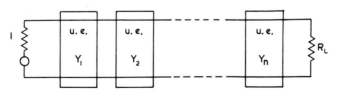

Figure 5.4.2   Pure unit element prototype

D. Equiripple without Inverters

$$Y_r = \begin{matrix} g_r & r \text{ odd} \\ 1/g_r & r \text{ even} \end{matrix} \qquad (5.4.10)$$

For $r = 1 \to n$

$$g_r = A_r \left[\frac{2\sin[(2r-1)\pi/2n]}{\alpha} - \frac{\alpha}{4} \left\{ \frac{\eta^2 + \sin^2(r\pi/n)}{\sin[(2r+1)\pi/2n]} \right. \right.$$
$$\left. \left. + \frac{\eta^2 + \sin^2[(r-1)\pi/n]}{\sin[(2r-3)\pi/2n]} \right\} \right] \qquad (5.4.11)$$

$$A_r = \frac{\{\eta^2 + \sin^2[(r-2)\pi/n]\}\{\eta^2 + \sin^2[(r-4)\pi/n]\}\ldots}{\{\eta^2 + \sin^2[(r-1)\pi/n]\}\{\eta^2 + \sin^2[(r-3)\pi/n]\}\ldots} \qquad (5.4.12)$$

where the last term $\eta^2 + \sin^2(0)$ is replaced by η, e.g.

$$A_2 = \frac{\eta}{\eta^2 + \sin^2(\pi/n)} \qquad (5.4.13)$$

and

$$R_L = 1 \qquad n \text{ odd}$$

or

$$R_L = \frac{\sqrt{1+\epsilon^2} - \epsilon}{\sqrt{1+\epsilon^2} + \epsilon} \qquad (5.4.14)$$

$$= \tanh^2\left[\frac{n}{2}\sinh^{-1}\eta\right] \qquad n \text{ even} \qquad (5.4.15)$$

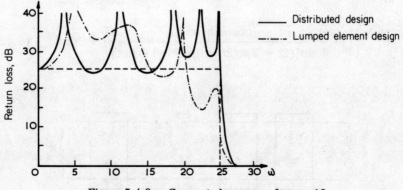

Figure 5.4.3   Computed response for $n = 10$, $\omega_0 = 25°$, $L_R = 25$ dB

To illustrate the accuracy of these design formulas the response characteristic for $n = 10$, $\omega_0 = 25°$ and a minimum return loss of 25 dB in the passband is plotted in Figure 5.4.3. To emphasize the passband behaviour, the return loss function $L_R = -20 \log |S_{11}|$ has been plotted against angular frequency. This result indicates the accuracy of these design formulas which are certainly within the tolerance one can achieve in physically building most devices based on this prototype. Furthermore, additional elements produced in most physical devices such as the lumped capacitance discontinuity in coaxial realizations may readily be taken into account by modifying the design equations and consequently, in general, greater accuracy in the design formulas is not required.

5.5 INTERDIGITAL FILTERS WITH MAXIMALLY FLAT AND CHEBYSHEV RESPONSE CHARACTERISTICS

For an optimum response characteristic of a network which contains n u.e.s and m stubs, all elements must contribute to the response. The maximally flat low-pass response will then be of the form

$$|S_{12}|^2 = \frac{1}{1 + F_n^2} \tag{5.5.1}$$

where

$$F_n = \left(\frac{\sin \omega}{\sin \omega_0}\right)^{m_1} \left(\frac{\tan \omega}{\tan \omega_0}\right)^{m_2} \tag{5.5.2}$$

and then the first $2(m_1 + m_2) - 1$ derivatives of $|S_{12}|^2$ will vanish at the origin.

For the corresponding equiripple solution, a particular generalized rational Chebyshev function must be used which may be expressed in the form (see Section A.1)

$$F_n = \epsilon \cosh\left(m_1 \cosh^{-1} \frac{\sin \omega}{\sin \omega_0} + m_2 \cosh^{-1} \frac{\tan \omega}{\tan \omega_0}\right) \tag{5.5.3}$$

The quasi band-pass response characteristics are obtained by replacing ω by $\tfrac{1}{2}\pi - \omega$ and for the case where $m_2 = 1$,

$$F_n = \left(\frac{\cot \omega}{\cot \omega_0}\right) \left(\frac{\cos \omega}{\cos \omega_0}\right)^{m_1} \tag{5.5.4}$$

for a maximally flat response and

$$F_n = \epsilon \cosh\left(\cosh^{-1} \frac{\cot \omega}{\cot \omega_0} + m_1 \cosh^{-1} \frac{\cos \omega}{\cos \omega_0}\right) \tag{5.5.5}$$

for an equiripple response.

For arbitrary values of m_1 and m_2, it is impossible to obtain a closed-form solution to the Hurwitz factorization problem ($m_1 = m_2$ being one exception, but it has little physical significance). However, adopting the process used in the previous section of expressing element values as a power-series expansion in the bandwidth scaling factor, there is no obvious necessity for possessing such a factorization as we shall now see.

Attention will be given entirely to the quasi band-pass case where $m_2 = 1$ and $m_1 = n - 1$ and the response characteristics described by (5.5.4) and (5.5.5). This class of characteristics are of the form necessary for realization by an n-wire interdigital filter and is significant from the physical realization viewpoint. It consists of a single n-wire line where coupling only exists between adjacent lines and alternate lines are short-circuited to ground at opposite ends. The other ends of the lines are open-circuited except for the first and last lines which provide the input and output ports. The device is illustrated in Figure 5.5.1. The structure is defined by its characteristic admittance matrix $[Y]$ which is of the form

$$[Y] = \begin{bmatrix} (Y_1 + Y_{12}) & -Y_{12} & 0 & & & \\ -Y_{12} & (Y_2 + Y_{12} + Y_{23}) & -Y_{23} & & & \\ 0 & -Y_{23} & (Y_3 + Y_{23} + Y_{34}) & & & \\ & & 0 & & & \\ & & 0 & & & \\ & & -Y_{34} & -Y_{n-1,n} & & \\ & & & Y_n + Y_{n-1,n} & & \end{bmatrix} \quad (5.5.6)$$

where $Y_{r,r+1}$ is the coupling admittance between the rth and $(r+1)$th lines and Y_r is the direct admittance to ground of the rth line. For this interdigital configuration it may be shown that apart from 1—1 transformers, which have no effect on any design process, the equivalent circuit is as shown in Figure 5.5.2.

Analysis of this circuit reveals that there is only a single transmission

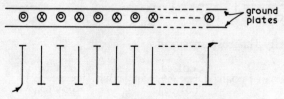

Figure 5.5.1 The interdigital filter

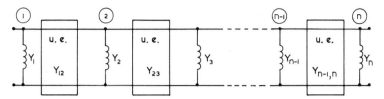

Figure 5.5.2 Equivalent circuit for interdigital filter

zero at the origin and the remaining transmission zeros are produced by the $n-1$ unit elements. Therefore for the maximally flat response.

$$|S_{12}|^2 = \frac{1}{1 + F_n^2} \tag{5.5.7}$$

where

$$F_n = \left(\frac{\cot \omega}{\cot \omega_0}\right) \left(\frac{\cos \omega}{\cos \omega_0}\right)^{n-1} \tag{5.5.8}$$

and for the Chebyshev case

$$F_n = \epsilon \cosh\left[\cosh^{-1}\frac{\cot \omega}{\cot \omega_0} + (n-1)\cosh^{-1}\frac{\cos \omega}{\cos \omega_0}\right] \tag{5.5.9}$$

5.6 EXPLICIT FORMULAS FOR ELEMENT VALUES IN CHEBYSHEV INTERDIGITAL FILTERS

From equations (5.5.7) and (5.5.9) for the Chebyshev interdigital filter we have

$$|S_{12}|^2 = \frac{1}{1 + \epsilon^2 H_n^2} \tag{5.6.1}$$

with

$$H_n = \cosh\left[\cosh^{-1}\left(\frac{\sqrt{1-\alpha^2}}{\alpha}\cot \omega\right) + (n-1)\cosh^{-1}\frac{\cos \omega}{\alpha}\right] \tag{5.6.2}$$

where $\alpha = \cos \omega_0$. For α small we have

$$H_n \approx \cosh\left(n \cosh^{-1}\frac{\cos \omega}{\beta}\right) \tag{5.6.3}$$

with

$$n \cos^{-1}\frac{x}{\beta} \approx \cos^{-1}\left(\frac{\sqrt{1-\alpha^2}}{\alpha}\frac{x}{\sqrt{1-x^2}}\right) + (n-1)\cos^{-1}\frac{x}{\alpha} \tag{5.6.4}$$

for $x \ll 1$ α small. Since

$$\cos^{-1} \eta = \frac{\pi}{2} - \eta + \frac{\eta^3}{6} - \ldots \tag{5.6.5}$$

$$\frac{1}{\beta} = \frac{1}{\alpha} - \frac{\alpha}{2n} + O(\alpha^3) \tag{5.6.6}$$

Therefore we have

$$|S_{12}|^2 \approx \frac{1}{1 + \epsilon^2 T_n^2(\cos \omega/\beta)} \tag{5.6.7}$$

where

$$\frac{1}{\beta} = \frac{1}{\alpha} - \frac{\alpha}{2n} + O(\alpha^3) \tag{5.6.8}$$

and

$$\alpha = \cos \omega_0 \tag{5.6.9}$$

which enables us to use the results on the explicit formulas for the basic stepped impedance distributed prototype filter.

Comparing equations (5.6.7) with (5.2.4) we note that $\omega \to \frac{1}{2}\pi + \omega$ and α is replaced by β. A complete basic section of the prototype comprises of a u.e. followed by an impedance inverter with an overall transfer matrix

$$\begin{bmatrix} \cos \omega & jZ_r \sin \omega \\ jY_r \sin \omega & \cos \omega \end{bmatrix} \begin{bmatrix} 0 & \dfrac{j}{K_{r,r+1}} \\ jK_{r,r+1} & 0 \end{bmatrix}$$

$$= \begin{bmatrix} -Z_r K_{r,r+1} \sin \omega & \dfrac{j \cos \omega}{K_{r,r+1}} \\ jK_{r,r+1} \cos \omega & \dfrac{-Y_r}{K_{r,r+1}} \sin \omega \end{bmatrix} \tag{5.6.10}$$

where Y_r and $K_{r,r+1}$ are given in equations (5.4.8) and (5.4.6). Replacing ω by $\frac{1}{2}\pi + \omega$ gives a transfer matrix

$$-\begin{bmatrix} Z_r K_{r,r+1} \cos \omega & \dfrac{j \sin \omega}{K_{r,r+1}} \\ jK_{r,r+1} \sin \omega & \dfrac{Y_r}{K_{r,r+1}} \cos \omega \end{bmatrix} \tag{5.6.11}$$

which, apart from 1—1 transformer may be identified with a u.e. of characteristic admittance

$$Y_{r,r+1} = K_{r,r+1} \tag{5.6.12}$$

with short-circuited shunt stubs of characteristic admittances Y_r'' and Y_{r+1}' which define an overall transfer matrix

$$
= \begin{bmatrix} 1 & 0 \\ -jY_r'' \cot \omega & 1 \end{bmatrix} \begin{bmatrix} \cos \omega & \dfrac{j \sin \omega}{K_{r,r+1}} \\ jK_{r,r+1} \sin \omega & \cos \omega \end{bmatrix} \begin{bmatrix} 1 & 0 \\ -jY_{r+1}' \cot \omega & 1 \end{bmatrix}
$$

$$
= \begin{bmatrix} \left(1 + \dfrac{Y_{r+1}'}{K_{r,r+1}}\right) \cos \omega & \dfrac{j \sin \omega}{K_{r,r+1}} \\ j\left[K_{r,r+1} \sin \omega - \dfrac{\cos^2 \omega}{\sin \omega}\left[Y_{r+1}' + Y_r''\left(1 + \dfrac{Y_{r+1}'}{K_{r,r+1}}\right)\right]\right] & \left(1 + \dfrac{Y_r''}{K_{r,r+1}}\right) \cos \omega \end{bmatrix}
$$

$$\tag{5.6.13}$$

Comparing (5.6.11) with (5.6.13) gives

$$Y_r'' = Y_r - K_{r,r+1} \tag{5.6.14}$$

and

$$Y_{r+1}' = Z_r K_{r,r+1}^2 - K_{r,r+1} \tag{5.6.15}$$

Thus, if we define the characteristic admittance matrix of the interdigital filter as

$$[Y] = \begin{bmatrix} Y_{11} - Y_{12} & 0 & 0 & & \\ -Y_{12} & Y_{22} & -Y_{23} & 0 & \\ 0 & -Y_{23} & Y_{33} & -Y_{34} & 0 \\ & & & & -Y_{n-1,n} \\ 0 & & & & Y_{n,n} \end{bmatrix} \tag{5.6.16}$$

then

$$Y_{r,r+1} = K_{r,r+1} \tag{5.6.17}$$

and

$$Y_{11} = Y_1$$

$$= \dfrac{2 \sin(\pi/2n)}{\eta \beta} - \dfrac{\beta}{4\eta} \left\{ \dfrac{\eta^2 + \sin^2(r\pi/n)}{\sin[(2r+1)\pi/2n]} - \dfrac{\eta^2}{\sin(\pi/2n)} \right\} + O(\beta^3)$$

$$\tag{5.6.18}$$

Also, from (5.6.14) and (5.6.15)

$$Y_{r,r} = Y_r + Z_{r-1} K_{r-1,r}$$

$$= \frac{2 \sin[(2r-1)\pi/n]}{\eta \beta} - \frac{\beta}{4\eta} \left\{ \frac{\eta^2 + \sin^2(r\pi/n)}{\sin[(2r+1)\pi/2n]} \right.$$

$$\left. - \frac{\eta^2 + \sin^2[(r-1)\pi/n]}{\sin[(2r-3)\pi/2n]} \right\} + O(\beta^3) \tag{5.6.19}$$

for $r > 1$.

By making the above identification, an error must occur due to the fact that there is a single transmission zero in the interdigital filter at the origin. This error will be minimized if it is assumed to occur in the centre of the filter and this is achieved by insisting that the design equations provide a symmetrical realization. Hence, equation (5.6.19) is allowed to apply up to the central element and the remaining elements are obtained from the symmetry condition.

5.7 SUMMARY OF RESULTS FOR CHEBYSHEV AND MAXIMALLY FLAT INTERDIGITAL FILTERS

Defining the interdigital filter with n digits by its characteristic admittance matrix

$$[Y] = \begin{bmatrix} Y_{11} & -Y_{12} & 0 & 0 & \\ -Y_{12} & Y_{22} & -Y_{23} & 0 & \\ 0 & -Y_{23} & Y_{23} & -Y_{34} & \\ 0 & & & & -Y_{n-1,n} \\ & & & & Y_{n,n} \end{bmatrix} \tag{5.7.1}$$

operating into 1 Ω terminations we have the following design equations.

A. Maximally Flat Response

$$Y_{r,r+1} = 1 \tag{5.7.2}$$

$$Y_{r,r} = Y_{n+1-r,\, n+1-r} = \frac{2 \sin[(2r-1)\pi/2n]}{\alpha}$$

$$- \frac{\alpha}{4} \left[\frac{1}{\sin[(2r+1)\pi/2n]} - \frac{1}{|\sin[(2r-3)\pi/2n]|} \right] + O(\alpha^3) \tag{5.7.3}$$

$$r = 1 \to \left[\frac{n+1}{2}\right], \quad [\ ] = \text{integer part of}$$

B. Chebyshev Response

For $r = 1 \to n - 1$

$$Y_{r,r+1} = \frac{\sqrt{\eta^2 + \sin^2(r\pi/n)}}{\eta} \tag{5.7.4}$$

For $r = 1 \to \left[\dfrac{n+1}{2}\right]$

$$Y_{r,r} = Y_{n+1-r,\,n+1-r} = \frac{2\sin[(2r-1)\pi/2n]}{\eta}\left[\frac{1}{\alpha} - \frac{\alpha}{2n}\right]$$
$$-\frac{\alpha}{4\eta}\left\{\frac{\eta^2 + \sin^2(r\pi/n)}{\sin[(2r+1)\pi/2n]} - \frac{\eta^2 + \sin^2[(r-1)\pi/n]}{|\sin[(2r-3)\pi/2n]|}\right\} + O(\alpha^3) \tag{5.7.5}$$

$\alpha = \cos \omega_0$

$$\eta = \sinh\left(\frac{1}{n}\sinh^{-1}\frac{1}{\epsilon}\right) \tag{5.7.6}$$

As an example of the accuracy of these design equations the response of a 17th-degree filter with $\omega_0 = 45°$ (3:1 band) and minimum value of passband return loss 21 dB, is shown in Figure 5.7.1. This is close to the maximum bandwidth one can attain with a physical realizable structure and exhibits a response very close to the exact equiripple one.

It is possible to scale the admittance level of the internal nodes of the filter without changing the response characteristic. This is achieved by

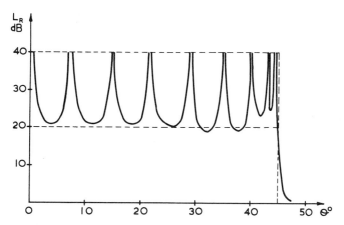

Figure 5.7.1 Computed response of interdigital filter designed from explicit formulas ($n = 17$, $\omega_0 = 45°$, $L_R = 21$ dB), $\omega = \pi/2 - \theta$

scaling the rth row and column of the admittance matrix by a constant within the constraint of a positive value for the direct admittance to ground of any line. Furthermore, for narrow bandwidths when α is small, it is necessary to introduce redundant lines $(0, n+1)$ at the input and output which represent u.e.s. of unity characteristic admittance. The new characteristic admittance matrix will then be

$$[Y] = \begin{bmatrix} 1 & -1 & 0 & 0 & & & 0 \\ -1 & 1+Y_{11} & -Y_{12} & 0 & & & \\ 0 & -Y_{12} & Y_{22} & -Y_{23} & & & \\ & & & & & -Y_{n-1,n} & 0 \\ 0 & & & & & (1+Y_{nn}) & -1 \\ 0 & & & & & -1 & 1 \end{bmatrix} \quad (5.7.7)$$

and for narrow bandwidths becomes

$$[Y] = \begin{bmatrix} 1 & -1 & 0 & & 0 \\ -1 & 1+\dfrac{2\sin(\pi/2n)}{\eta\alpha} & \dfrac{-\sqrt{\eta^2+\sin^2(\pi/n)}}{\eta} & & 0 \\ 0 & \dfrac{-\sqrt{\eta^2+\sin^2(\pi/n)}}{\eta} & \dfrac{2\sin(3\pi/2n)}{\eta\alpha} & & \dfrac{-\sqrt{\eta^2+\sin^2(2\pi/n)}}{\eta} \end{bmatrix}$$

(5.7.8)

for a Chebyshev response. If the self-admittances to ground are required to be of the order of unity, then internal scaling yields,

$$[Y] = \begin{bmatrix} 1 & -\sqrt{\dfrac{\eta\alpha}{2\sin(\pi/2n)}} & & \\ \sqrt{\dfrac{\eta\alpha}{2\sin(\pi/2n)}} & 1+\dfrac{\eta\alpha}{2\sin(\pi/2n)} & & \\ 0 & -\dfrac{\alpha}{2}\dfrac{\sqrt{\eta^2+\sin^2(\pi/n)}}{\sqrt{\sin(\pi/2n)\sin(3\pi/2n)}} & & \\ 0 & & & 0 \\ -\dfrac{\alpha}{2}\dfrac{\sqrt{\eta^2+\sin^2(\pi/n)}}{\sqrt{\sin(\pi/2n)\sin(3\pi/2n)}} & & & \\ 1 & & -\dfrac{\alpha}{2}\dfrac{\sqrt{\eta^2+\sin^2(2\pi/n)}}{\sqrt{\sin(3\pi/2n)\sin(\pi/2n)}} \end{bmatrix}$$

(5.7.9)

which may be summarized as:

$$Y_{r,r} = 1 \qquad r = 0, 2 \to n-1, n+1 \tag{5.7.10}$$

$$Y_{1,1} = Y_{n,n} = 1 + \frac{\eta\alpha}{2\sin(\pi/2n)} \tag{5.7.11}$$

$$Y_{0,1} = Y_{n-1,n} = \sqrt{\frac{\eta\alpha}{2\sin(\pi/2n)}} \tag{5.7.12}$$

$$Y_{r,r+1} = \frac{\alpha}{2}\sqrt{\frac{\eta^2 + \sin^2(r\pi/n)}{\sin[(2r-1)\pi/2n]\sin[(2r+1)\pi/2n]}} \qquad r = 1 \to n-1 \tag{5.7.13}$$

5.8 FOURIER COEFFICIENT DESIGN TECHNIQUE FOR STEPPED IMPEDANCE TRANSMISSION LINE FILTERS

The design formulas which have been given in the previous sections for distributed prototype filters have only been approximate. In this section an algorithm will be developed for evaluating the element values in a cascade of u.e.s without recourse to formal synthesis procedures and thus partially retaining the accuracy of direct explicit form solutions. Furthermore, the technique is not restricted to the maximally flat and Chebyshev cases, but allows an arbitrary characteristic to be approximated in a prescribed sense. The function which will be approximated is the return loss L_R given by

$$L_R = -20\log|S_{11}(j\tan\omega)| \geq 0 \tag{5.8.1}$$

which, since it is even and periodic, may be written as a Fourier series,

$$L_R = \frac{a_0}{2} + \sum_{r=1}^{\infty} a_r \cos 2r\omega \tag{5.8.2}$$

which is valid at all points where the function and its derivatives are continuous. Initially we shall consider the value of these Fourier coefficients in the maximally flat and Chebyshev cases.

From equation (5.2.9), for the maximally flat case, $S_{11}(z)$ may be written as,

$$S_{11}(z) = \prod_{r=1}^{n} \left\{ \frac{(1-z)\exp[\sinh^{-1}(j\alpha e^{j\theta_r})]}{1 - z\exp[2\sinh^{-1}(j\alpha e^{j\theta_r})]} \right\} \qquad \theta_r = \frac{(2r-1)\pi}{2n} \tag{5.8.3}$$

and for $|\alpha| < 1$,

$$\ln[S_{11}(z)] = n\ln(1-z) + \sum_{r=1}^{n} [\sinh^{-1}(j\alpha e^{j\theta_r})$$

$$-\ln\{1 - z\exp[2\sinh^{-1}(j\alpha e^{j\theta_r})]\}]$$

$$= \frac{a_0}{2} + \sum_{m=1}^{\infty} a_m z^m \qquad |z| \leq 1 \qquad (5.8.4)$$

where

$$a_m = \frac{(-1)^m \sum_{r=1}^{n} \{\exp[2m\sinh^{-1}(j\alpha e^{j\theta_r})] - n\}}{m} \qquad (5.8.5)$$

However,

$$e^{q\sinh^{-1}x} = 1 + qx + \frac{q^2 x^2}{2!} + \frac{q(q^2-1)x^3}{3!}$$

$$+ \frac{q^2(q^2-4)x^4}{4!} + \frac{q(q^2-1)(q^2-9)x^5}{5!}$$

$$+ \frac{q^2(q^2-4)(q^2-16)x^6}{6!} + \ldots \qquad (5.8.6)$$

and

$$\sum_{r=1}^{n} j\exp[jq\theta_r] = \begin{matrix} 0 & q \text{ even}, n>1 \\ \\ -\frac{1}{\sin(q\pi/2n)} & q \text{ odd} \end{matrix} \qquad (5.8.7)$$

Hence,

$$a_m = (-1)^{m+1} 2\left[\frac{\alpha}{\sin(\pi/2n)} - \frac{(4m^2-1)\alpha^3}{3!\sin(3\pi/2n)}\right.$$

$$\left. + \frac{(4m^2-1)(4m^2-9)\alpha^5}{5!\sin(5\pi/2n)} + \ldots \ m \leq n \right. \qquad (5.8.8)$$

From (5.8.4)

$$\ln\left[S_{11}\left(\frac{1}{z}\right)\right] = \frac{a_0}{2} + \sum_{m=1}^{\infty} a_m z^{-m} \qquad |z| \geq 1 \qquad (5.8.9)$$

and hence

$$\frac{L_R}{8.68} = \frac{1}{2}\left\{\ln\left[S_{11}(z)\right] + \ln\left[S_{11}\left(\frac{1}{z}\right)\right]\right\}_{z=e^{-j2\omega}}$$

$$= \frac{a_0}{2} + \sum_{m=1}^{\infty} a_m \cos 2m\omega \qquad (5.8.10)$$

which is the Fourier series representation of the return loss function.
For the Chebyshev case, equation (5.2.20) gives

$$S_{11}(z) = \frac{F(z,0)}{F(z,\eta)} \qquad (5.8.11)$$

with

$$F(z,\eta) = \sum_{r=1}^{n}\left[\frac{1 - ze^{-2\sinh^{-1}(j\alpha\cos\psi_r)}}{e^{-\sinh^{-1}(j\alpha\cos\psi_r)}}\right] \qquad (5.8.12)$$

$$\psi_r = \sin^{-1} j\eta + \frac{(2r-1)\pi}{2n}$$

For $|\alpha|<1$

$$\ln[S_{11}(z)] = A(z,0) \qquad (5.8.13)$$

with

$$A(z,\eta) = -\sum_{r=1}^{n}[\sinh^{-1}(j\alpha\cos\psi_r) - \ln(1 - ze^{-2\sinh^{-1}(j\alpha\cos\psi_r)})]$$

$$= \frac{a_0}{2} + \sum_{m=1}^{\infty} a_m z^m \qquad (5.8.14)$$

where

$$a_m = (-1)^m \frac{\left[\sum_{r=1}^{m}(e^{-m\sinh^{-1}(j\alpha\cos\psi_r)} - e^{-2m\sinh^{-1}(j\alpha\cos\theta_r)})\right]}{m} \qquad (5.8.15)$$

Now it may be shown that

$$\sum_{r=1}^{n}[(\cos\psi_r)^j - (\cos\theta_r)^j] = \begin{array}{l} 0 \\ H_j \end{array} \begin{array}{l} j \text{ even} \\ j \text{ odd} \end{array} \qquad (5.8.16)$$

where

$$H_{2i+1} = \sum_{q=1}^{i+1} \frac{(2i+1)!!(i!)^2}{(2i)!!(i+q)!(i-q+1)!} \cdot \frac{\sinh[(2q-1)\sinh^{-1}\eta]}{\sin[(2q-1)\pi/2n]}$$

$$(5.8.17)$$

Hence,

$$a_m = (-1)^{m+1} 2 \left\{ \frac{\alpha\eta}{\sin(\pi/2n)} - \frac{(4m^2-1)}{3!} \right.$$
$$\left[\frac{\eta}{\sin(\pi/2n)} + \frac{\sinh(3\sinh^{-1}\eta)}{\sin(3\pi/2n)} \right] \frac{\alpha^3}{4} + \frac{(4m^2-1)4m^2-9)}{5!}$$
$$\left[\frac{10\eta}{\sin(\pi/2n)} + \frac{5\sinh(3\sinh^{-1}\eta)}{\sin(3\pi/2n)} + \frac{\sinh(5\sinh^{-1}\eta)}{\sin(5\pi/2n)} \right] \frac{\alpha^5}{16}$$
$$\left. - \ldots \right\} \qquad (5.8.18)$$

$$= (-1)^{m+1} 2 \left\{ \sum_{i=1}^{\infty} \frac{\alpha^{2i-1}(-4)^{i-1}\left(m+i-\frac{3}{2}\right)!}{(2i-1)!\left(m-i+\frac{1}{2}\right)!} \right.$$
$$\left. \sum_{q=1}^{i} K_q \frac{\sinh[(2q-1)\sinh^{-1}\eta]}{\sin[(2q-1)\pi/2n]} \right\} \qquad m \leqslant n \qquad (5.8.19)$$

where

$$K_q = \frac{(2i-1)!![(i-1)!]^2}{(2i-2)!!(i+q-1)!(i-q)!}$$

An alternative method of approximating the ideal filter response shown in Figure 5.8.1 is to immediately form the Fourier series.

For

$$\begin{aligned} L_R &= A & \omega < \omega_0 \\ & 0 & \omega > \omega_0 \end{aligned} \qquad (5.8.21)$$

in the basic half-period we have the Fourier series

$$L_R = \frac{2A\omega_0}{\pi} \left[1 + \sum_{r=1}^{\infty} \frac{\sin 2r\omega_0}{r\omega_0} \cos 2r\omega \right] \qquad (5.8.22)$$

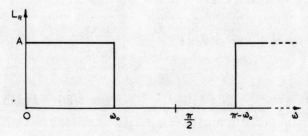

Figure 5.8.1 Ideal distributed filter response

This particular solution will be discussed further later in this section.

In general, we shall be interested in the filter comprising a cascade of u.e.s whose return loss

$$\frac{L'_R}{8.68} = \frac{a_0'}{2} + \sum_{r=1}^{\infty} a_r' \cos 2r\omega \qquad (5.8.23)$$

is such that

$$a_r' = a_r \qquad \text{for } r = 0 \rightarrow n \qquad (5.8.24)$$

Initially we must establish some general theorems[5.2].

Theorem 1

Let the function $F(\theta)$ possess a Fourier series representation, valid almost everywhere, in the form

$$F(\theta) = \frac{\gamma_0}{2} + \sum_{r=1}^{\infty} \gamma_r \cos 2r\theta \qquad (5.8.25)$$

then if

$$F(\theta) \geq 0 \qquad (5.8.26)$$

the sequence $\{\gamma_i\}_n = \{\gamma_0, \gamma_1, \gamma_2, \ldots, \gamma_n\}$ is a positive definite sequence for all n; i.e.

$$\sum_{r=0}^{n} \sum_{m=0}^{n} \gamma_{r-m} E_r E_m^* \geq 0 \qquad (\gamma_{-q} = \gamma_q) \qquad (5.8.27)$$

for any arbitrary set of constants E_q.

Proof

Since $F(\theta) \geq 0$ and the Fourier coefficients of this particular cosine series are given by

$$\gamma_r = \frac{4}{\pi} \int_0^{\frac{1}{2}\pi} F(\theta) \cos 2r\theta \, d\theta \qquad r = 0 \rightarrow n \qquad (5.8.28)$$

we have

$$\int_{-\frac{1}{2}\pi}^{\frac{1}{2}\pi} \left| \sum_{r=0}^{n} E_r e^{j2r\theta} \right|^2 F(\theta) \, d\theta \geq 0 \qquad (5.8.29)$$

i.e.

$$\frac{\pi}{2} \sum_{r=0}^{n} \sum_{m=0}^{n} \gamma_{r-m} E_r E_m^* \geq 0 \qquad (\gamma_{-q} = \gamma_q) \qquad (5.8.30)$$

since

$$\int_{-\frac{1}{2}\pi}^{\frac{1}{2}\pi} F(\theta) \sin 2r\theta \, d\theta = 0 \qquad (5.8.31)$$

It should also be noted that the statement 'the sequence $\{\gamma_i\}_n$ is a positive definite sequence' is equivalent to the statement that the matrix

$$[\gamma]_n = \begin{bmatrix} \gamma_0 & \gamma_1 & \gamma_2 & & \gamma_n \\ \gamma_1 & \gamma_0 & \gamma_1 & & \gamma_{n-1} \\ & & \gamma_0 & & \gamma_1 \\ \gamma_n & & & \gamma_1 & \gamma_0 \end{bmatrix} \qquad (5.8.32)$$

is a positive definite matrix, i.e.

$$[E^+][\gamma]_n[E] \geqslant 0 \qquad (5.8.33)$$

for all constant n-vectors $[E]$ where $[E^+]$ is the transpose conjugate.

This particular theorem may immediately be applied to the return loss function L_R which must be non-negative thus enforcing the Fourier coefficients a_r to form a positive definite sequence $\{\gamma_i\}_n$ for all n.

Theorem 2
If the sequence $\{\gamma_i\}_n$ is a positive definite sequence, then the sequence $\{c_i\}_n$ is also positive definite if

$$\frac{c_0}{2} + \sum_{r=1}^{\infty} c_r z^r = L\left[\frac{\gamma_0}{2} + \sum_{r=1}^{\infty} \gamma_r z^r + H(z)\right] \qquad |z| < 1 \qquad (5.8.34)$$

where $L(s)$ and $H(z)$ are any positive real operators of the form

$$\begin{aligned}&\operatorname{Re} L(s) \geqslant 0 \\ &L(s) \text{ is real for } s \text{ real}\end{aligned} \qquad \operatorname{Re} s \geqslant 0 \qquad (5.8.35)$$

$$\begin{aligned}&\operatorname{Re} H(z) \geqslant 0 \\ &H(z) \text{ is real for } z \text{ real}\end{aligned} \qquad |z| \leqslant 1 \qquad (5.8.36)$$

Proof
Any series whose coefficients form a positive definite sequence may be expressed as[5.3]

$$\frac{\gamma_0}{2} + \sum_{r=1}^{\infty} \gamma_r z^r = \frac{2}{\pi} \int_0^{\frac{1}{2}\pi} \left(\frac{1 + z e^{j2\theta}}{1 - z e^{j2\theta}}\right) F(\theta) d\theta \qquad |z| < 1 \qquad (5.8.37)$$

where $F(\theta)$ is given in (5.8.25) and (5.8.26).

If
$$\phi(z) = \frac{\gamma_0}{2} + \sum_{r=1}^{\infty} \gamma_r z^r \qquad |z| < 1 \qquad (5.8.38)$$

then
$$\operatorname{Re} \phi(z) = \frac{2}{\pi} \int_0^{\frac{1}{2}\pi} \frac{1-a^2}{1 - 2a\cos(2\theta + \psi) + a^2} F(\theta) d\theta \qquad z = ae^{j\psi}$$
$$(5.8.39)$$

and hence
$$\operatorname{Re} \phi(z) \geqslant 0 \qquad \text{for } |z| < 1 \qquad (5.8.40)$$

Let
$$C(z) = L[\phi(z) + H(z)] \qquad (5.8.41)$$

where $L(s)$ and $H(z)$ are positive real operators described by equations (5.8.35) and (5.8.36), then

$$\operatorname{Re} C(z) \geqslant 0$$
$$C(z) \text{ is real for } z \text{ real} \qquad |z| < 1 \qquad (5.8.42)$$

But
$$\operatorname{Re} C(e^{j2\theta}) = \frac{c_0}{2} + \sum_{r=1}^{\infty} c_i \cos(2r\theta) \qquad \text{almost everywhere} \qquad (5.8.43)$$

and from Theorem 1, $\{c_i\}_n$ forms a positive definite sequence.
We are now in a position to establish the main theorem:

Theorem 3
Let the desired return loss function of a network be defined as
$$\frac{L_R}{8.68} = -\ln |S_{11}| \geqslant 0 \qquad (5.8.44)$$

It may be expressed almost everywhere at real frequencies by the Fourier series
$$\frac{L_R}{8.68} = \frac{a_0}{2} + \sum_{r=1}^{\infty} a_r \cos 2r\omega \qquad (5.8.45)$$

and the input impedance $Z(z)$ of a network which, when operating into a 1 Ω generator produces this return loss function L_R is given by
$$Z(z) = 1 + 2 \sum_{r=1}^{\infty} c_r z^r \qquad |z| < 1 \qquad (5.8.46)$$

where the $\{c_i\}_n$ form a positive definite sequence and are obtained through the following algorithm

$$b_r = \frac{1}{r!} \begin{vmatrix} a_1 & -1 & 0 & 0 & & & 0 \\ 2a_2 & a_1 & -2 & 0 & & & 0 \\ 3a_3 & 2a_2 & a_1 & -3 & & & \\ & & & & (r-2) & & 0 \\ & & & & & a_1 & -(r-1) \\ ra_r & (r-1)a_{r-1} & & & & 2a_2 & a_1 \end{vmatrix} \quad (5.8.47)$$

$$c_1 = e^{-\frac{1}{2}a_0} \tag{5.8.48}$$

$$c_{r+1} = (-1)^r c_1 \begin{vmatrix} b_1 - c_1 & b_2 & b_3 & & b_r \\ 1 & b_1 - c_1 & b_2 & & b_{r-1} \\ 0 & 1 & b_1 - c_1 & & \\ & & 1 & b_1 - c_1 & b_2 \\ 0 & & 0 & 1 & b_1 - c_1 \end{vmatrix}$$

$$(5.8.49)$$

Proof

Given the return loss function $L_R \geqslant 0$, we express it as

$$L_R = \frac{a_0}{2} + \sum_{r=1}^{\infty} a_r \cos 2r\omega \quad \text{(a.e.)} \tag{5.8.50}$$

by computing the Fourier coefficients

$$a_r = \frac{4}{\pi} \int_0^{\frac{1}{2}\pi} L_R \cos 2r\omega \, d\omega \tag{5.8.51}$$

with $\{a_i\}_n$ forming a positive definite sequence. Thus

$$\frac{L_R}{8.68} = -\ln |S_{11}(e^{j\omega})| = \frac{a_0}{2} + \sum_{r=1}^{\infty} a_r \cos 2r\omega \tag{5.8.52}$$

For the case where $S_{11}(z)$ is analytic in $|z| \leqslant 1$ and devoid of zeros in $|z| < 1$ we have

$$-\ln S_{11}(z) = \frac{a_0}{2} + \sum_{r=1}^{\infty} a_r z^r \tag{5.8.53}$$

for $|z| < 1$ and almost everywhere on $|z| = 1$, or

$$\frac{1}{S_{11}(z)} = \exp\left(\frac{1}{2}a_0 + \sum_{r=1}^{\infty} a_r z^r\right) \tag{5.8.54}$$

$$= e^{\frac{1}{2}a_0}\left(1 + \sum_{r=1}^{\infty} b_r z^r\right) \qquad (5.8.55)$$

Equating (5.8.54) to (5.8.55) leads to a set of linear simultaneous equations relating the b_i to the a_i whose solution may be expressed as

$$b_r = \frac{1}{r!}\begin{vmatrix} a_1 & -1 & 0 & 0 & & 0 \\ 2a_2 & a_1 & -2 & 0 & & 0 \\ 3a_3 & 2a_2 & a_1 & -3 & & \\ & & & a_1 & & -(r-1) \\ rq_r & (r-1)a_{r-1} & & 2a_2 & & a_1 \end{vmatrix} \qquad (5.8.56)$$

Introducing a redundant u.e. of unity characteristic impedance by forming the new non-minimum phase reflection coefficient $zS_{11}(z)$, we have the input impedance

$$Z(z) = \frac{1 + zS_{11}(z)}{1 - zS_{11}(z)} = \frac{1 + \sum_{r=1}^{\infty} b_r z^r + ze^{-\frac{1}{2}a_0}}{1 + \sum_{r=1}^{\infty} b_r z^r - ze^{-\frac{1}{2}a_0}} \qquad (5.8.57)$$

$$= 1 + 2\sum_{r=1}^{\infty} c_r z^r \qquad (5.8.58)$$

Equating (5.8.57) and (5.8.58) again leads to a set of simultaneous equations with the solutions

$$c_1 = e^{-\frac{1}{2}a_0} \qquad (5.8.59)$$

and

$$c_{r+1} = (-1)^r c_r \begin{vmatrix} b_1 - c_1 & b_2 & b_3 & & b_r \\ 1 & b_1 - c_1 & b_2 & & b_{r-1} \\ 0 & 1 & b_1 - c_1 & & \\ & & 1 & b_1 - c_1 & b_2 \\ 0 & & 0 & 1 & b_1 - c_1 \end{vmatrix}$$
$$(5.8.60)$$

From equations (5.8.54) and (5.8.58) we have

$$Z(z) = \tanh[\phi(z) - \ln z] \qquad (5.8.61)$$

where

$$\phi(z) = \frac{a_0}{2} + \sum_{r=1}^{\infty} a_r z^r \qquad |z| < 1 \qquad (5.8.62)$$

From Theorem 1, since $L_R \geq 0$ the sequence $\{a_1\}_n$ is positive definite. From Theorem 2, Re $\phi(z) \geq 0$ for $|z| < 1$ and since Re$(-\ln z) \geq 0$ for $|z| < 1$ and tanh( ) is a positive real operator, Re $Z(z) \geq 0$ for $|z| < 1$ and $Z(0) = 1$. Thus,

$$Z(z) = 1 + 2 \sum_{r=1}^{\infty} c_r z^r \tag{5.8.63}$$

is a positive real function and $\{c_i\}_n$ is a positive definite sequence.

Finally, it follows that if a network possesses an input impedance

$$Z'(z) = 1 + 2 \sum_{r=1}^{\infty} c'_r z^r \tag{5.8.64}$$

where

$$c'_r = c_r \qquad r = 1 \to n \tag{5.8.65}$$

then the return loss function L'_R of this network will be of the form

$$\frac{L'_R}{8.68} = \frac{a'_0}{2} + \sum_{r=1}^{\infty} a'_r z^r \tag{5.8.66}$$

where

$$a'_r = a_r \qquad r = 0 \to n-1 \tag{5.8.67}$$

that is, the first n Fourier coefficients of the return loss function will be equal to the corresponding coefficients in the desired return loss function.

We shall now proceed to determine an explicit expression for the element values in a cascade of u.e.s from a given impedance function expressed as a power series in z.

5.9 EXPLICIT FORMULAS FOR ELEMENT VALUES IN ARBITRARY STEPPED IMPEDANCE TRANSMISSION LINE FILTERS

In this section we shall be concerned with determining the values for the characteristic impedance of a cascade of u.e.s whose input impedance is given in the form

$$Z(z) = 1 + 2 \sum_{r=1}^{\infty} c_r a^r \qquad |z| < 1 \tag{5.9.1}$$

where $\{c_i\}_n$ is a positive definite sequence. If $Z(z)$ is the input impedance of a cascade of n u.e.s it may be written in the form

$$Z(z) = \frac{Q_n(z)}{P_n(z)} \quad (5.9.2)$$

with

$$P_n(0) = 1 = Q_n(0) \quad (5.9.3)$$

since $Z(0) = 1$ and

$$P_n \tilde{Q}_n + \tilde{P}_n Q_n = K_n z^n \quad (5.9.4)$$

where

$$\begin{aligned} \tilde{P}_n &= z^n P_n(z^{-1}) \\ \tilde{Q}_n &= z^n Q_n(z^{-1}) \end{aligned} \quad (5.9.5)$$

If Q_n and P_n satisfy (5.9.4), then it may be shown that[5.4]

$$P_n(z) = \frac{1}{\Delta_{n-1}} \begin{vmatrix} 1 & c_1 & c_2 & c_n \\ c_1 & 1 & c_1 \\ c_{n-1} & c_{n-2} & & c_1 \\ z^n & z^{n-2} & & 1 \end{vmatrix} \quad (5.9.6)$$

where

$$\Delta_{n-1} = \begin{vmatrix} 1 & c_1 & c_2 & c_{n-1} \\ c_1 & 1 & c_2 & \\ & & & \\ c_{n-1} & & & 1 \end{vmatrix} \quad (5.9.7)$$

and

$$\frac{Q_n(z)}{P_n(z)} = 1 + 2 \sum_{r=1}^{\infty} c_r z^r \quad |z| < 1 \quad (5.9.8)$$

Obviously if two polynomials $P_m(z)$ and $Q_m(z)$, $m \leq n$ are chosen such that

$$P_m \tilde{Q}_m + \tilde{P}_m Q_m = K_m z^m \quad (5.9.9)$$

$$P_m(0) = Q_m(0) = 1 \quad (5.9.10)$$

and

$$\frac{Q_m(z)}{P_m(z)} = 1 + 2 \sum_{r=1}^{\infty} c'_r z^r \quad (5.9.11)$$

where

$$c'_r = c_r \quad r = 1 \to m \quad (5.9.12)$$

then

$$P_m(z) = \frac{1}{\Delta_{m-1}} \begin{vmatrix} 1 & c_1 & c_2 & c_m \\ c_1 & 1 & c_1 & \\ c_{m-1} & c_{m-2} & & c_1 \\ z^m & z^{m-1} & & 1 \end{vmatrix} \quad (5.9.13)$$

Thus, we have a sequence of n polynomials $\{P_i(z)\}_n$ which are defined by (5.9.13). It is possible to obtain a recurrence formula for these polynomials directly using (5.9.13). However, we shall immediately investigate the network significance of (5.9.12) which will automatically produce this recurrence formula.

Consider a cascade of u.e.s of unity characteristic impedance and separated by ideal transformers. If the input impedance $Z(z)$ is formed when this network is terminated in a $1\,\Omega$ resistor, then the transfer matrix of the network must be of the form:

$$[T_n] = \frac{1}{2K_n z^{\frac{1}{2}n}} \begin{bmatrix} Q_n + \tilde{Q}_n & Q_n - \tilde{Q}_n \\ P_n - \tilde{P}_n & P_n + \tilde{P}_n \end{bmatrix} \quad (5.9.14)$$

A transformer of turns ratio t_n may now be extracted from the output side followed by the extraction of a u.e. of unity characteristic impedance to yield a remaining matrix

$$[T_{n-1}] = [T_n] \begin{bmatrix} t_n & 0 \\ 0 & \dfrac{1}{t_n} \end{bmatrix} \frac{1}{2z^{1/2}} \begin{bmatrix} z+1 & z-1 \\ z-1 & z+1 \end{bmatrix}$$

$$= \frac{1}{2K_{n-1} z^{\frac{1}{2}(n-1)}} \begin{bmatrix} Q_{n-1} + \tilde{Q}_{n-1} & Q_{n-1} - \tilde{Q}_{n-1} \\ P_{n-1} - \tilde{P}_{n-1} & P_{n-1} + \tilde{P}_{n-1} \end{bmatrix} \quad (5.9.15)$$

where

$$\frac{zP_{n-1}}{K_{n-1}} = \frac{z}{2K_n}\left(t_n + \frac{1}{t_n}\right)P_n + \frac{z}{2K_n}\left(\frac{1}{t_n} - t_n\right)\tilde{P}_n$$

or, since $P_{n-1}(0) = 1$,

$$(1 - \gamma_n^2)P_{n-1} = P_n - \gamma_n \tilde{P}_n \quad (5.9.16)$$

where

$$\gamma_n = \left.\frac{P_n}{\tilde{P}_n}\right|_{z=\infty} = \frac{t_n^2 - 1}{t_n^2 + 1} \quad (5.9.17)$$

Furthermore, from (5.9.14) and (5.9.15),

$$Z_{n-1}(z) = Z_n(z) + z^{n-1}\left(\sum_{r=0}^{\infty} d_r z^r\right) \quad (5.9.18)$$

and the first $n-1$ Fourier coefficients are identical. Thus, $P_{n-1}(z)$ is described by (5.9.13) with $m = n-1$.

We may now proceed with this synthesis process and generate the sequence of polynomials $\{P_i(z)\}_n$ which must satisfy the recurrence formula

$$(1 - \gamma_m^2)P_{m-1} = P_m - \gamma_m \tilde{P}_m \qquad (5.9.19)$$

with

$$\gamma_m = \left.\frac{P_m}{\tilde{P}_m}\right|_{z=\infty} \qquad (5.9.20)$$

for $m = 1 \to n$ and where

$$t_m = \left(\frac{1+\gamma_m}{1-\gamma_m}\right)^{1/2} \qquad (5.9.21)$$

is the value of the transformer between the $(m-1)$th and mth u.e.s of unity characteristic impedance. If the transformers are taken to the output of the network then a cascade of u.e.s will be formed where the characteristic impedances are given by

$$Z_{r+1} = Z_r \frac{1+\gamma_r}{1-\gamma_r} \qquad (5.9.22)$$

and the γ_r is identifiable as the junction reflection coefficients.

Finally, from (5.9.13) and (5.9.20)

$$\gamma_m = \frac{(-1)^m}{\Delta_{m-1}} \begin{vmatrix} c_1 & c_2 & & c_m \\ 1 & c_1 & c_2 & \\ c_1 & 1 & c_1 & \\ c_{m-2} & & 1 & c_1 \end{vmatrix} \qquad (5.9.23)$$

for $m = 1 \to n$ establishing, with equation (5.9.22), the explicit formula for the characteristic impedances of a cascade of u.e.s with input impedance

$$Z(z) = 1 + 2 \sum_{r=1}^{\infty} c_r z^r \qquad (5.9.24)$$

Thus, using the algorithms given in the last two sections, the element values for the u.e.s may be obtained directly from the Fourier coefficients of the return loss function. One particular realization which is important is the symmetrical one. If a symmetrical network of degree $2n-1$ is designed such that the first n Fourier coefficients are correct then the numerator $N_q(z)$ of the reflection coefficient $S_{11}(z)$ must

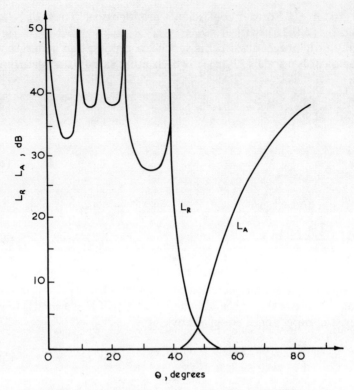

Figure 5.9.1 Computed response of symmetrical filter designed from Fourier coefficients ($n = 9$, $\omega_0 = 45°$, $A = 30$ dB)

possess the property

$$z^q N_q(z^{-1}) = \pm N_q(z) \qquad (5.9.25)$$

and since $S_{11}(z)$ is minimum phase, $N_q(z) \neq 0$ for $|z| < 1$. From (5.9.25), $N_q(z) \neq 0$ for $|z| > 1$ and therefore the zeros of $S_{11}(z)$ are all at real frequencies. This may be demonstrated by taking the response given in (5.8.22) and realizing the first n terms with a symmetrical network of degree $2n - 1$. For a degree-9 network with $\omega_0 = 45°$ and $A = 30$ dB the result shown in Figure 5.9.1 may be obtained showing a close similarity to the Chebyshev case defined in equations (5.8.19) and (5.8.20).

CHAPTER 6

Phase Approximations for Distributed Networks

6.1 INTRODUCTION

As in the case of lumped networks, phase approximations for finite rational transfer functions reduce to phase approximations for polynomials. However, in the distributed case the variable is $t = \tanh p$ instead of p and this makes a significant difference to the linear phase approximation.

Consider the polynomial $P(x)$; then

$$\psi_\ell = \text{Arg } P(j\omega) = \tan^{-1}\left[\frac{O(\omega)}{E(\omega)}\right] \tag{6.1.1}$$

where

$$P(j\omega) = E(\omega) + jO(\omega) \tag{6.1.2}$$

For the distributed case,

$$\psi_d = \text{Arg } P(j \tan \omega) = \tan^{-1}\left[\frac{O(\tan \omega)}{E(\tan \omega)}\right] \tag{6.1.3}$$

and any linear phase approximation in (6.1.3) will not correspond to a linear phase approximation in (6.1.1). In terms of the group delay response, for lumped networks

$$T_{g\ell}(j\omega) = \frac{d\psi_\ell}{d\omega} \tag{6.1.4}$$

and for the distributed case

$$T_{gd}(j\omega) = \frac{d\psi_d}{d\omega} = \frac{d\psi_d}{d(\tan \omega)} \frac{d(\tan \omega)}{d\omega}$$

$$= T_{g\ell}(j \tan \omega)(1 + \tan^2 \omega)$$

$$= \frac{T_{g\ell}(j \tan \omega)}{\cos^2 \omega} \tag{6.1.5}$$

which readily demonstrates that the constant group delay approximation differs from the lumped case.

One immediate result of this difference is that it is possible to obtain an exact linear phase or constant delay at all frequencies from an nth-degree polynomial and this is discussed in the next section. However, using these polynomials for combined amplitude and phase approximations leads to difficulties in meeting amplitude requirements and this is discussed in Chapter 7. Since a linear phase response for a filter is only required over the passband region it is advantageous to restrict the approximation to linear phase to occur only over a finite band. For this case a maximally flat solution is obtained and then a finite band equidistant solution.

6.2 EXACT LINEAR PHASE POLYNOMIALS

For an exact linear phase response we require

$$\psi(\omega) = \alpha\omega \tag{6.2.1}$$

where

$$\psi(\omega) = \operatorname{Arg} P_n(j \tan \omega) \tag{6.2.2}$$

Thus,

$$\frac{P_n(-j \tan \omega)}{P_n(j \tan \omega)} = e^{-2\alpha\omega} \tag{6.2.3}$$

or

$$\frac{P_n(-t)}{P_n(t)} = \left(\frac{1-t}{1+t}\right)^\alpha \tag{6.2.4}$$

and for $P_n(t)$ to be an nth-degree polynomial $\alpha = n$ and

$$P_n(t) = (1 + t)^n \tag{6.2.5}$$

with a group delay

$$T_g = n \tag{6.2.6}$$

6.3 MAXIMALLY FLAT DISTRIBUTED LINEAR PHASE POLYNOMIAL

For a maximally flat approximation around the origin ($t = 0$) to a linear phase response we require

$$\psi(\omega) = \alpha\omega + a_1 \omega^{2n+1} + a_2 \omega^{2n+3} + \ldots \tag{6.3.1}$$

where

$$\psi(\omega) = \text{Arg } A_n^{(\alpha)}(j \tan \omega) \tag{6.3.2}$$

and the polynomial $A_n^{(\alpha)}(t)$ will be dependent upon the delay α at the origin.

Thus,

$$\frac{A_n^{(\alpha)}(-j \tan \omega)}{A_n^{(\alpha)}(j \tan \omega)} = e^{-2\alpha\omega} e^{-2a_1 \omega^{2n+1} + \ldots} \tag{6.3.3}$$

or

$$A_n^{(\alpha)}(t)(1-t)^\alpha - A_n^{(\alpha)}(-t)(1+t)^\alpha = b_1 t^{2n+1} + b_2 t^{2n+3} + \ldots \tag{6.3.4}$$

which uniquely defines $A_n^{(\alpha)}(t)$ within a constant multiplier. As in the lumped case discussed in Chapter 3 there are several methods of obtaining this polynomial.[6.1] However, we shall adopt the definite integral representation approach, and normalizing the constant term to unity, prove that $A_n^{(\alpha)}(t)$ may be expressed as

$$A_n^{(\alpha)}(t) = \frac{t^{2n+1} \Gamma(\alpha+n+1)}{(1-t)^\alpha \Gamma(2n+1)\Gamma(\alpha-n)} \int_1^{1/t} (1-xt)^{\alpha-n-1}(x^2-1)^n \, dx$$

$$(\alpha > n) \quad (6.3.5)$$

where $\Gamma(x)$ is the Gamma function.

Initially we shall show that $A_n^{(\alpha)}(t)$ defined by equation (6.3.5) is indeed an nth-degree polynomial by developing the degree-varying recurrence formula.

Let

$$I_{n+1} = \int_1^{1/t} (1-xt)^{\alpha-n-2}(x^2-1)^{n+1} \, dx \tag{6.3.6}$$

and integrating by parts gives

$$I_{n+1} = \left[\frac{-(1-xt)^{\alpha-n-1}(x^2-1)^{n+1}}{(\alpha-n-1)t} \right]_1^{1/t}$$

$$+ \frac{2(n+1)}{(\alpha-n-1)t} \int_1^{1/t} (1-xt)^{\alpha-n-1} x(x^2-1)^n \, dx$$

$$= \frac{2(n+1)}{(\alpha-n-1)t} \int_1^{1/t} (1-xt)^{\alpha-n-1} x(x^2-1)^n \, dx \tag{6.3.7}$$

for $\alpha > n+1$, $n \geq 0$. Integrating by parts again gives

$$I_{n+1} = \frac{2(n+1)}{(\alpha-n)(\alpha-n-1)t^2} \int_1^{1/t} (1-xt)^{\alpha-n}[(x^2-1)^n + 2nx^2(x^2-1)^{n-1}]dx$$

$$= \frac{2(n+1)}{(\alpha-n)(\alpha-n-1)t^2} \int_1^{1/t} (1-xt)^{\alpha-n}[(2n+1)(x^2-1)^n + 2n(x^2-1)^{n-1}]dx$$

$$= \frac{2(n+1)}{(\alpha-n)(\alpha-n-1)t^2} \left[(2n+1)I_n - \frac{t^2(\alpha-n-1)}{2(n+1)}I_{n+1} + 2nI_{n-1} \right] \quad (6.3.8)$$

or

$$\frac{\alpha-n+1}{\alpha-n}I_{n+1} = \frac{2(n+1)}{(\alpha-n)(\alpha-n-1)t^2}[(2n+1)I_n + 2nI_{n-1}]$$

$$(6.3.9)$$

From (6.3.5) and (6.3.6), this recurrence formula reduces to

$$A_{n+1}^{(\alpha)}(t) = A_n^{(\alpha)}(t) + \frac{t^2(\alpha^2-n^2)}{4n^2-1}A_{n-1}^{(\alpha)}(t) \quad (6.3.10)$$

Direct evaluation of (6.3.5) for $n = 0, 1$ gives

$$A_0^{(\alpha)}(t) = 1, \qquad A_1^{(\alpha)}(t) = 1 + \alpha t \quad (6.3.11)$$

which are the initial conditions for the degree-varying recurrence formula (6.3.10) establishing that $A_n^{(\alpha)}(t)$ is an nth-degree polynomial in t.

A differential recurrence formula may also be obtained from the definite integral representation of $A_n^{(\alpha)}(t)$. From (6.3.5)

$$\frac{d[A_n^{(\alpha)}(t)(1-t)^\alpha]}{dt} = (1-t)^{\alpha-1}\left[(1-t)\frac{dA_n^{(\alpha)}(t)}{dt} - \alpha A_n^{(\alpha)}(t)\right]$$

$$= \frac{t^{2n}\Gamma(\alpha+n+1)}{\Gamma(2n+1)\Gamma(\alpha-n)}[(2n+1)I_n + tI_n'] \quad (6.3.12)$$

where

$$I_n' = (\alpha-n-1)\int_1^{1/t}(1-xt)^{\alpha-n-2}x(x^2-1)^n dx \quad (6.3.13)$$

$$= -(\alpha-n-1)\int_1^{1/t}(1-xt)^{\alpha-n-2}$$

$$\frac{[t(x^2-1)+(t+x)(1-xt)]}{1-t^2}(x^2-1)^n\,dx$$

$$= -(\alpha-n-1)\left\{\frac{tI_{n+1}+tI_n+[(\alpha-n-1)t/2(n+1)]I_{n+1}}{1-t^2}\right\}$$

(6.3.14)

using equations (6.3.6) and (6.3.7).
Thus,

$$\frac{d[A_n^{(\alpha)}(t)(1-t)^{\alpha}]}{dt} = \frac{t^{2n}\Gamma(\alpha+n+1)\{[(2n+1)-(\alpha+n)t^2]I_n - t^2[(\alpha+n+1)(\alpha-n-1)/2(n+1)]I_{n+1}\}}{\Gamma(2n+1)\Gamma(\alpha-n)(1-t^2)}$$

(6.3.15)

or

$$\frac{dA_n^{(\alpha)}(t)}{dt} - \frac{\alpha}{(1-t)}A_n^{(\alpha)}(t)$$

$$= \frac{1}{(1-t^2)t}\{[(2n+1)-(\alpha+n)t^2]A_n^{(\alpha)}(t) - (2n+1)A_{n+1}^{(\alpha)}(t)\}$$

(6.3.16)

and using the recurrence formula (6.3.10) we finally have

$$(1-t^2)\frac{dA_n^{(\alpha)}(t)}{dt} = (\alpha-nt)A_n^{(\alpha)}(t) - \frac{t(\alpha^2-n^2)}{(2n-1)}A_{n-1}^{(\alpha)}(t) \quad (6.3.17)$$

which is the differential recurrence formula.

Several other recurrence formulas may be developed with regard to shifting the parameter α by integer quantities. One useful result is to relate $A_n^{(\alpha)}(t)$, $A_{n+1}^{(\alpha)}(t)$ and $A_n^{(\alpha+1)}(t)$ as follows.
From (6.3.6)

$$I_n^{(\alpha+1)} = \int_1^{1/t}(1-xt)^{\alpha-n}(x^2-1)^n\,dx$$

$$= \int_1^{1/t}(1-xt)^{\alpha-n-1}[(x^2-1)^n - xt(x^2-1)^n]\,dx$$

$$= I_n^{(\alpha)} - \frac{t^2(\alpha-n-1)}{2(n+1)}I_{n+1}^{(\alpha)} \qquad (6.3.18)$$

using (6.3.7). Hence,

$$(\alpha - n)(1 - t)A_n^{(\alpha+1)}(t) = (\alpha + n + 1)A_n^{(\alpha)}(t) - (2n + 1)A_{n+1}^{(\alpha)}(t) \qquad (6.3.19)$$

or using (6.3.10)

$$(1 - t)A_n^{(\alpha+1)}(t) = A_n^{(\alpha)}(t) - \frac{t^2(\alpha + n)}{2n - 1}A_{n-1}^{(\alpha)}(t) \qquad (6.3.20)$$

We may now use equations (6.3.10) and (6.3.17) to prove that $A_n^{(\alpha)}(t)$ is indeed the maximally flat distributed linear phase polynomial by proving that the associated group delay is a maximally flat approximation to the constant value α around $t = 0$. Now

$$\begin{aligned} T_g &= (1 - t^2)\operatorname{Ev}\left[\frac{dA_n^{(\alpha)}(t)}{dt}\bigg/A_n^{(\alpha)}(t)\right] \\ &= \alpha - \frac{\alpha^2 - n^2}{2n - 1}\operatorname{Ev}\left[\frac{tA_{n-1}^{(\alpha)}(t)}{A_n^{(\alpha)}(t)}\right] \\ &= \alpha - \frac{t(\alpha^2 - n^2)[A_{n-1}^{(\alpha)}(t)A_n^{(\alpha)}(-t) - A_{n-1}^{(\alpha)}(-t)A_n^{(\alpha)}(t)]}{(2n - 1)A_n^{(\alpha)}(t)A_n^{(\alpha)}(-t)} \end{aligned} \qquad (6.3.21)$$

where use has been made of (6.3.17). From the degree-varying recurrence formula (6.3.10)

$$\operatorname{Od}[A_{n+1}^{(\alpha)}(t)A_n^{(\alpha)}(-t)] = \frac{t^2(\alpha^2 - n^2)}{4n^2 - 1}\operatorname{Od}[A_n^{(\alpha)}(t)A_{n-1}^{(\alpha)}(t)]$$

$$= K_n t^{2n+1} \qquad (6.3.22)$$

by repeated application. Thus combining (6.3.22) with (6.3.21), and noticing that T_g must be zero at $t = 1$, we have

$$T_g = \alpha\left[1 - \frac{A_n^{(\alpha)}(1)A_n^{(\alpha)}(-1)}{A_n^{(\alpha)}(t)A_n^{(\alpha)}(-t)}\right] \qquad (6.3.23)$$

which proves that $A_n^{(\alpha)}(t)$ is the required polynomial.

As in the lumped case, it is important to determine the conditions under which $A_n^{(\alpha)}(t)$ is a Hurwitz polynomial in t. Forming the admittance function

$$Y_r^{(\alpha)}(t) = \frac{(2r - 1)A_r^{(\alpha)}(t)}{\alpha t A_{r-1}^{(\alpha)}(t)} \qquad (6.3.24)$$

and substituting into (6.3.10) gives,

$$Y_{r+1}^{(\alpha)}(t) = \frac{2r + 1}{\alpha t} + \frac{K_{r,r+1}^2}{Y_r^{(\alpha)}(t)} \qquad (6.3.25)$$

where

$$K_{r,r+1} = \frac{\sqrt{\alpha^2 - r^2}}{\alpha} \tag{6.3.26}$$

Now if $Y_r^{(\alpha)}(t)$ is a positive real function then $Y_{r+1}^{(\alpha)}(t)$ will also be a positive real function if $K_{r,r+1}$ is real, that is

$$\alpha \geq r - 1 \tag{6.3.27}$$

Thus, in a similar manner to the lumped case, $A_n^{(\alpha)}(t)$ is a Hurwitz polynomial if

$$\alpha \geq n - 1 \tag{6.3.28}$$

and the network interpretation of the recurrence formula (6.3.25) is shown in Figure 6.3.1 in terms of short-circuited shunt stubs and ideal impedance inverters of characteristic admittance $K_{m,m+1}$.

To complete the section on the maximally flat distributed linear phase polynomial it may be observed that the polynomial, known as the symmetrical Jacobi polynomial[6.2]

$$Q_n^{(\alpha)}(t) = t^n A_n^{(\alpha)}(1/t) \tag{6.3.29}$$

is the polynomial which gives rise to a maximally flat approximation to a linear phase response around $t = \infty$. This may be established by noting that any linear transformation in frequency does not alter the form of the approximation. In particular, applying the frequency transformation

$$\omega \to \omega + \frac{\pi}{2} \tag{6.3.30}$$

we have

$$\tan \omega \to -\frac{1}{\tan \omega} \tag{6.3.31}$$

or in the complex domain,

$$t \to 1/t \tag{6.3.32}$$

Thus, we immediately have the necessary results to design quasi band-pass linear phase responses, a result which has no counterpart in the lumped domain.

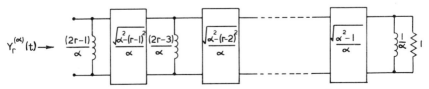

Figure 6.3.1   Network interpretation of $Y_r^{(\alpha)}(t)$

Finally, two further properties which immediately follow are:
$$A_n^{(n)}(t) = (1 + t)^n \tag{6.3.33}$$
recovering the exact linear phase polynomial and
$$A_n^{(\alpha)}\left(\frac{p}{\alpha}\right)\bigg|_{\alpha=\infty} = P_n(p) \tag{6.3.34}$$
where $P_n(p)$ is the Bessel polynomial.

6.4 EQUIDISTANT AND ARBITRARY DISTRIBUTED LINEAR PHASE POLYNOMIALS

As in the lumped case, one finite band approximation to the linear phase response is the equidistant approximation defined with respect to the nth-degree polynomial $A_n^{(\alpha)}(t)$

$$\text{Arg } A_n^{(\alpha)}(\pm j \tan r\theta_0 \mid \theta_0) = \pm \alpha_r \theta_0 \qquad r = 0 \rightarrow n \tag{6.4.1}$$

The associated phase error function

$$\epsilon(\omega) = \alpha\omega - \psi(\tan \omega) \tag{6.4.2}$$

will have the behaviour in one period as shown in Figure 6.4.1. Unfortunately, a definite integral representation for this polynomial is not known[6.3] and therefore an approach similar to the one adopted in the lumped case for the arbitrary phase polynomials will be used. As this is the case, the formal solution to the distributed arbitrary phase polynomial will be developed specifically recovering the equidistant linear phase case with an explicit solution for the degree-varying recurrence formula.

Let $Q_n(t)$ be a polynomial in t and

$$\text{Arg } Q_n(\pm j\omega_r) = \pm j\psi_r \qquad r = 1 \rightarrow n \tag{6.4.3}$$

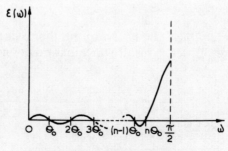

Figure 6.4.1 Phase response of equidistant distributed phase polynomial

Then following the process given in Section 3.5 we construct the function

$$F_n(t) = \frac{Q_n(t)}{\prod_{r=1}^{n}(1 + t^2/\omega_r^2)}$$

$$= \sum_{r=1}^{n} A_r^n \frac{(1 + B_r t)}{(1 + t^2/\omega_r^2)} \tag{6.4.4}$$

From condition (6.4.3)

$$B_r = \frac{\tan \psi_r}{\omega_r} \tag{6.4.5}$$

which is the same as equation (3.5.6) and consequently the A_r^n are given by equation (3.5.12). Thus, the sequence of polynomials $\{Q_m(t)\}$ defined by (6.4.3) satisfy a recurrence formula

$$Q_{n+1}(t) = \alpha_n Q_n(t) + (t^2 + \omega_n^2) Q_{n-1}(t) \tag{6.4.6}$$

if the leading coefficient is normalized to unity where (see equation 3.5.19)

$$\alpha_n = \left(1 - \frac{\omega_n^2}{\omega_{n+1}^2}\right) \frac{A_n^{n+1}}{A_n^n} \tag{6.4.7}$$

with the initial conditions

$$Q_0(t) = 1, \qquad Q(t) = t + \frac{\omega_1}{\tan \psi_1} \tag{6.4.8}$$

Similarly, an arbitrary phase polynomial of the second kind may be generated.

Returning to the equidistant linear phase polynomial where $\omega_r = \tan r\theta_0$ and $\psi_r = \alpha r\theta_0$, it is more convenient to normalize the constant term to unity to give

$$A_{n+1}^{(\alpha)}(t \mid \theta_0) = A_n^{(\alpha)}(t \mid \theta_0) + \beta_n(t^2 + \tan^2 n\theta_0) A_{n-1}^{(\alpha)}(t \mid \theta_0) \tag{6.4.9}$$

with the initial conditions

$$A_0^{(\alpha)}(t \mid \theta_0) = 1, \qquad A_1^{(\alpha)}(t \mid \theta_0) = 1 + \frac{\tan \alpha\theta_0}{\tan \theta_0} t \tag{6.4.10}$$

and it may be shown that [6.4]

$$\beta_n = \frac{1}{2} \frac{[\tan^2 \alpha\theta_0 - \tan^2 n\theta_0][\tan(n+1)\theta_0 - \tan(n-1)\theta_0] \tan \theta_0}{[\tan^2 (n+1)\theta_0 - \tan^2 n\theta_0][\tan^2 n\theta_0 - \tan^2 (n-1)\theta_0]} \tag{6.4.11}$$

which may be reduced to

$$\beta_n = \frac{\cos(n-1)\theta_0 \, \cos(n+1)\theta_0 \, \sin(\alpha+n)\theta_0 \, \sin(\alpha-n)\theta_0 \, \cos^2 n\theta_0}{\sin(2n-1)\theta_0 \, \sin(2n+1)\theta_0 \, \cos^2 \alpha\theta_0}$$
(6.4.12)

The proof of this result may be summarized as initially establishing that $A_n^{(\alpha)}(t \mid \theta_0)$ is a polynomial of degree n in $\tan \alpha\theta_0$ as well as t; $A_n^{(\alpha)}(t \mid \theta_0) = A_{n-1}^{(\alpha)}(t \mid \theta_0)$ when $\alpha = \pm n$ thus determining the leading factor of (6.4.11), and finally evaluating at $\alpha\theta_0 = \frac{1}{2}\pi - \theta_0$ to determine the dependence of β_n on $\tan \theta_0$.

The sequence of polynomials $\{A_i^{(\alpha)}(t\theta_0)\}_n$ will form a Hurwitz sequence if

$$\beta_r \geqslant 0 \qquad r = 0 \to n-1 \tag{6.4.13}$$

which may be obtained in a manner similar to that employed in Chapter 3. Hence from (6.4.12), $A_n^{(\alpha)}(t \mid \theta_0)$ will be Hurwitz if

$$\alpha \geqslant n-1, \qquad n\theta_0 \leqslant \frac{\pi}{2}, \qquad (\alpha+n-1)\theta_0 \leqslant \pi \tag{6.4.14}$$

6.5 DISTRIBUTED ALL-PASS AND REFLECTION FILTERS

The methods employed in this section closely follow those used in Section 3.7 and hence only the relevant differences will be emphasized. The all-pass transfer function in the distributed domain may be written as

$$S_{12}(t) = \frac{H_q(-t)}{H_q(t)} \tag{6.5.1}$$

and for a maximally flat solution to constant delay around $t = 0$, we have

$$H_q(t) = A_n^{(\alpha)}(t)(1+t)^m \tag{6.5.2}$$

where $A_n^{(\alpha)}(t)$ is the maximally flat distributed linear phase polynomial. The input admittance of the corresponding reflection filter is

$$Y_q'(t) = \frac{H_q(t) + H_q(-t)}{H_q(t) - H_q(-t)} \tag{6.5.3}$$

which may be synthesized as a ladder-type network containing n stubs and m unit elements. If the m unit elements are initially extracted, then they will all be of unity impedance due to the form of (6.5.2) and after extracting them all the remaining admittance will be

$$Y_n'(t) = \frac{A_n^{(\alpha)}(t) + A_n^{(\alpha)}(-t)}{A_n^{(\alpha)}(t) - A_n^{(\alpha)}(-t)} \tag{6.5.4}$$

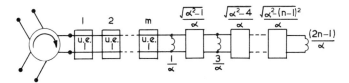

Figure 6.5.1   Realization of $Y_n(t)$ as a stub filter

for m even and the reciprocal for m odd. Viewed from the output end assuming a 1-Ω termination at the input, regardless of the m unit elements, the admittance will be

$$Y_n(t) = \frac{(2n-1)A_n^{(\alpha)}(t)}{\alpha t\, A_{n-1}^{(\alpha)}(t)} \tag{6.5.5}$$

and follows directly from Section 3.7 due to the form of the recurrence formula (6.3.10). Synthesis of this admittance as a ladder-type network with stubs separated by impedance inverters produces the network shown in Figure 6.5.1 from Section 6.3.

For $m = 0$, we have a stub filter consisting of shunt short-circuited and series open-circuited stubs once the inverters have been transformed out of the network. However, for practical reasons it would be preferable to have $n-1$ unit elements and transform them into the network in order to separate each stub by a unit element. Thus, the synthesis of $Y_n(t)$ given in (6.5.5) should be in the form of a stub-unit element extraction process. From (6.3.24), (6.3.25) and (6.3.26) we have shown that if

$$Y_r^{(\alpha)}(t) = \frac{(2r-1)A_r^{(\alpha)}(t)}{\alpha t A_{r-1}^{(\alpha)}(t)} \tag{6.5.6}$$

then the extraction of a stub of admittance

$$\frac{2r-1}{\alpha t} \tag{6.5.7}$$

followed by an impedance inverter $K_{r-1,r}$ leaves a remaining admittance

$$Y_{r-1}^{(\alpha)}(t) \tag{6.5.8}$$

where

$$K_{r-1,r} = \frac{\sqrt{\alpha^2 - (r-1)^2}}{\alpha} \tag{6.5.9}$$

Extraction of a u.e. from an admittance $Y_r^{(\alpha)}(t)$ leaves a remaining admittance

$$Y'(t) = \frac{Y_r^{(\alpha)}(t) - tY_r^{(\alpha)}(1)}{1 - tZ_r^{(\alpha)}(1)Y_r^{(\alpha)}(t)} \tag{6.5.10}$$

where $Y_r^{(\alpha)}(1)$ is the characteristic admittance of the extracted u.e. Using (6.5.12) this reduces to

$$Y'(t) = \frac{2r-1}{\alpha t} \left\{ \frac{A_r^{(\alpha)}(t) - t^2 [A_r^{(\alpha)}(1)/A_{r-1}^{(\alpha)}(1)] A_{r-1}^{(\alpha)}(t)}{A_{r-1}^{(\alpha)}(t) - [A_{r-1}^{(\alpha)}(1)/A_r^{(\alpha)}(1)] A_r^{(\alpha)}(t)} \right\} \qquad (6.5.11)$$

which from (6.3.19) and (6.3.20) becomes

$$Y'(t) = \frac{2r-1}{\alpha t} \frac{A_r^{(\alpha+1)}(t)(\alpha + r)}{A_{r-1}^{(\alpha+1)}(t)(\alpha - r + 1)} \qquad (6.5.12)$$

Finally, extraction of a transformer with turns ratio

$$n_r = \sqrt{\frac{(\alpha + r)(\alpha + 1)}{(\alpha - r + 1)\alpha}} \qquad (6.5.13)$$

leaves the admittance

$$Y''(t) = Y_r^{(\alpha+1)}(t) \qquad (6.5.14)$$

and the synthesis cycle may be repeated.

After completing the entire synthesis, the transformers and inverters may be removed by transforming across to the output end to leave the network shown in Figure 6.5.2 where, for $r = 1 \to n - 1$,

$$g_{r+1} = \frac{(\alpha + n - 2r)(\alpha + n - 2r - 1)}{(4r^2 - 1)g_r} \qquad (6.5.15)$$

$$Y_{r+1} = \frac{\alpha + n - 2r - 1}{\alpha + n - 2r - 2Y_r}$$

with the initial conditions

$$g_1 = \alpha + n - 1$$

$$Y_1 = \frac{\alpha + n - 2}{\alpha + n - 1} \qquad (6.5.16)$$

The corresponding band-pass solution is obtained by replacing $t \to 1/t$ to give a reflection coefficient for the reflection filter of the form

$$S_{12}(t) = \frac{Q_n^{(\alpha)}(-t)(1-t)^{n-1}}{Q_n^{(\alpha)}(t)(1+t)^{n-1}} \qquad (6.5.17)$$

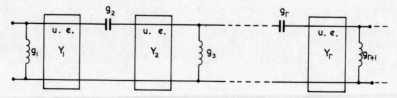

Figure 6.5.2   Stub-unit element realization of $Y_n(t)$ (low-pass)

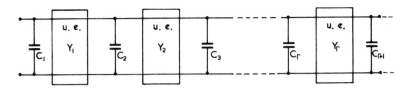

Figure 6.5.3 Stub-unit element realization of $Y_n(t)$ (band-pass)

and direct synthesis results in the network shown in Figure 6.5.3 where,
For $r = 1 \to n-1$,

$$C_{r+1} = \frac{(\alpha + n - 2r)(2r + 1)}{(\alpha + n - 2r - 1)(2r - 1)} C_r$$

$$Y_{r+1} = \frac{\alpha + n - 2r - 2}{\alpha + n - 2r - 1} Y_r \qquad (6.5.18)$$

with the initial conditions

$$C_1 = \frac{1}{\alpha + n - 1}, \qquad Y_1 = \frac{\alpha + n - 2}{\alpha + n - 1} \qquad (6.5.19)$$

These results may be obtained directly from the low-pass case if it is recognized that the transformation of $t \to 1/t$ for a u.e. results in a u.e. in cascade with an ideal impedance inverter.

Finally, although no explicit formulas for element values exist, reflection filters may be synthesized using finite-band arbitrary phase polynomials.

CHAPTER 7

Simultaneous Amplitude and Phase Approximations for Distributed Networks

7.1 INTRODUCTION

The types of transfer functions in which we shall be primarily interested are those which are capable of being realized by a resistively terminated reciprocal lossless two-port. In the low-pass case,

$$S_{12}(t) = \frac{E_{2n-1}(t)}{D_{2n}(t)} \quad \text{or} \quad \frac{E_{2n}(t)}{D_{2n}(t)} \tag{7.1.1}$$

or

$$S_{12}(t) = \frac{E_{2n}(t)}{D_{2n+1}(t)} \tag{7.1.2}$$

where $E_{2n-2}(t)$ and $E_{2n}(t)$ are even polynomials. For the band-pass case,

$$S_{12}(t) = \frac{\sqrt{1-t^2}\, O_{2n-1}(t)}{D_{2n}(t)} \tag{7.1.3}$$

or

$$S_{12}(t) = \frac{O_{2n+1}(t)}{D_{2n+1}(t)} \tag{7.1.4}$$

where $O_{2n-1}(t)$ and $O_{2n+1}(t)$ are odd polynomials. The form chosen for (7.1.3) is the one which may be realized directly in the form of a generalized interdigital filter.[7.1]

As shown in the last chapter, as distinct from the lumped case, in the distributed domain it is possible for transfer functions to exhibit constant delay at all frequencies. By enforcing this constraint, the form of the possible amplitude responses are obtained and approximations to the ideal amplitude response are discussed in the following section. The next two sections deal with phase equalization of filters designed on an amplitude basis and amplitude equalization of filters designed with a finite band linear phase response. The limitations of these approaches are overcome in the combined amplitude and phase approximations where the constraints cannot be considered independently. In the band-pass case particular emphasis is placed upon the form of response for realization by generalized interdigital filters and the appropriate synthesis processes are given.

7.2 CONSTANT AMPLITUDE FILTERS WITH EXACT LINEAR PHASE

The quasi low-pass transfer function with constant delay at all frequencies is

$$S_{12}(t) = \frac{E_m(t)}{(1+t)^n} \tag{7.2.1}$$

where $E_m(t)$ is an even polynomial of degree $n-1$, for n odd, and n or $n-2$, for n even. Thus

$$|S_{12}(j \tan \omega)|^2 = P_n^2(\cos \omega) \tag{7.2.2}$$

where $P_n(x)$ is an even or odd polynomial of degree n. Thus, the approximation problem reduces to determining the polynomial $P_n(x)$ which possess a two-band constraint of the form

$$\begin{aligned} 1 \geqslant P_n(x) \geqslant 1 - \epsilon_1^2 & \quad 1 \geqslant |x| \geqslant x_2 \\ |P_n(x)| \leqslant \epsilon_2^2 & \quad |x| \leqslant x_1 \end{aligned} \tag{7.2.3}$$

illustrated by the shaded region in Figure 7.2.1.

Consider a particular solution to the approximation problem which leads to an equiripple solution in the region $|x| \leqslant x_1$ with the optimum number of turning points as shown in Figure 7.2.2. Now the construction of a polynomial $Q_m(x)$ which satisfies the condition $|Q_m(x)| \leqslant \epsilon_2^2$ for $|x| \leqslant x_1$ and $1 \geqslant Q_m(x) > P_n(x)$ for some $|x| > x_1$ immediately

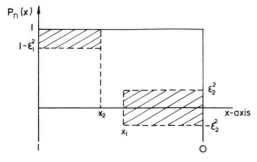

Figure 7.2.1   Specification of polynomial $P_n(x)$

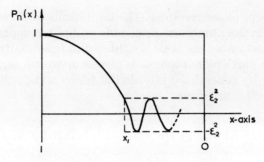

Figure 7.2.2 Equiripple solution in stop-band

results in $m > n$. Thus one might assume that this is the optimum solution to the approximation problem. However, this is not the case as may be illustrated by investigating the particular solution to the above problem, which is,[7.2]

$$P_n(x) = \frac{T_n(x/x_1)}{T_n(x/x_1)} \qquad (7.2.4)$$

where $T_n(x) = \cos(n \cos^{-1} x)$. Thus given x_1 and the ripple level, the degree is uniquely determined! Conversely, given x_1, the ripple level is reduced as n increases and $P_n(x)$ decreases in the passband as n increases. Thus, if a particular passband specification is given, and for any given n is not satisfied, then increasing n will never allow the specification to be met. This general problem arises from the fact that degree, ripple level and cut-off frequency are not independent qualities as distinct from the conventional Chebyshev response. One may place equal ripples in the passband instead of the stopband, e.g.

$$P_{2n}(x) = 1 - \epsilon_1^2 T_n^2\left(\frac{x}{x^2}\right) \qquad (7.2.5)$$

Although it is not absolutely necessary to have $P_{2n}(1) = 0$, $P_{2n}(1) \approx 0$ and therefore

$$\frac{1}{\epsilon_1^2} \approx T_n^2\left(\frac{1}{x_2}\right) \qquad (7.2.6)$$

and again degree and ripple level are related for a fixed cut-off frequency. These results are readily appreciated if the degenerate maximally flat solutions are recovered as

$$P_n(x) = x^n = \cos^n \omega \qquad (7.2.7)$$

from (7.2.4), or
$$P_{2n}(x) = 1 - x^{2n} = 1 - \cos^{2n} \omega \tag{7.2.8}$$
from (7.2.5).

Thus, it is essential to have an extra degree of freedom in order to meet any given two-band specification. This is achieved by placing a different number of ripples in each band for the equiripple solution or a different number of derivatives equal to zero in the passband and stopband for the maximally flat solution. No analytical solution exists for the equiripple case but it has been solved numerically.[7.3] By definition the maximally flat solution is

$$P_{2n+m-1}(\cos \omega) = \frac{\int_\omega^{\frac{1}{2}\pi} \sin^{2n-1} \theta \cos^m \theta \, d\theta}{\int_0^{\frac{1}{2}\pi} \sin^{2n-1} \theta \cos^m \theta \, d\theta} \tag{7.2.9}$$

Thus, expanding about $\omega = 0$,
$$P_{2n+m-1}(\cos \omega) = 1 - a\omega^{2n} + b\omega^{2n+2} + \ldots \tag{7.2.10}$$
and about $\omega = \frac{1}{2}\pi$
$$P_{2n+m-1}(\cos \omega) = c(-\omega + \tfrac{1}{2}\pi)^{m+1} + d(-\omega + \tfrac{1}{2}\pi)^{m+3} + \ldots \tag{7.2.11}$$

where
$$c = \frac{2na}{m+1} \tag{2.7.12}$$

illustrating how a specification may be met but not necessarily proving that any arbitrary specification may be met by a finite-degree filter. This, however, may be proved by considering a non-optimum solution to the approximation problem of the form[7.2]

$$|S_{12}|^2 = \frac{T_n^2[K\sqrt{1-\epsilon^2 T_m^2(\sin \omega/\sin \omega_0)}]}{T_n^2(K)} \tag{7.2.13}$$

Thus, although a finite-degree filter may be found to realize any filter specification, the degree, even when the optimum two-band equiripple solution is used, may be excessive. For example, to meet the selectivity of a conventional Chebyshev filter of degree 10, a filter of degree of the order of 200 would be required. Qualitatively this may readily be appreciated since a linear phase response is unnecessarily being provided in the stopband region and in the transition band. This, coupled with the fact that transfer functions of this form are very difficult to realize by practical structures, renders them of little use.

7.3 CONSTANT AMPLITUDE FILTERS WITH PHASE EQUALIZATION

The overall transfer function of a phase-equalized distributed filter may be written as

$$S_{12}(t) = S_{12}^{A}(t) \frac{G(-t)}{G(t)} \qquad (7.3.1)$$

where $S_{12}^{A}(t)$ produces the required amplitude response, and the all-pass factor the phase equalization as described in Section 4.2. The only difference between the lumped and distributed cases is the modification of the group delay or phase response, but this does not affect the type of solution. For example, in the Chebyshev case where

$$|S_{12}(j \tan \omega)|^2 = \frac{1}{1 + \epsilon^2 T_n^2(\tan \omega / \tan \omega_0)} \qquad (7.3.2)$$

$$\theta_n = -\sum_{r=1}^{n-1} \tan^{-1} \frac{\eta}{\sin(r\pi/n)}$$

$$\operatorname{Arg} S_{12}^{A}(-j[\tan \omega_0 \cos(2q+1)\pi/2n])$$

$$= \theta_n + 2 \sum_{r=1}^{q} \tan^{-1} \frac{\eta}{\sin(r\pi/n)} \qquad q = 0 \to n-1 \qquad (7.3.3)$$

from equation (4.2.16) and one may then obtain the arbitrary phase polynomial which equalizes the phase at these points of interpolation.

The degree considerations are the same as the lumped case, and in addition to this the physical complexity of the required equalizers does not make the technique particularly attractive.

7.4 LINEAR PHASE FILTERS WITH AMPLITUDE EQUALIZATION

Consider a transfer function of the form

$$S_{12}(t) = \frac{E_{2m}(t)}{A_n^{(\alpha)}(t)} \qquad (7.4.1)$$

where $A_n^{(\alpha)}(t)$ is the maximally flat distributed linear phase polynomial and hence

$$\operatorname{Arg} S_{12}(j \tan \omega) = -\alpha \omega + a_1 \omega^{2n+1} + \ldots \qquad (7.4.2)$$

as in Section 4.3. Thus, if the remaining constraints are applied to produce a maximally flat amplitude response around the origin, we require

$$\frac{E_{2m}(t)}{A_n^{(\alpha)}(t)} - \left(\frac{1-t}{1+t}\right)^{\frac{1}{2}\alpha} = b_1 t^{2m+1} + b_2 t^{2m+2} + \ldots \tag{7.4.3}$$

or $E_{2m}(t)/(1-t^2)^n$ is the first $2m$ terms of

$$\left(\frac{1-t}{1+t}\right)^{\frac{1}{2}\alpha} \frac{A_n^{(\alpha)}(t)}{(1-t^2)^n} \tag{7.4.4}$$

Representing $E_{2m}(t)$ as

$$\frac{E_{2m}(t)}{(1-t^2)^n} = \sum_{r=0}^{n} a_r \left(\frac{t^2}{1-t^2}\right)^r \tag{7.4.5}$$

we have $a_0 = 1$. a_1 is obtained by differentiating (7.4.4) and (7.4.5) with respect to t, multiplying by $(1-t^2)^2/t$ and equating. Differentiating (7.4.4) results in

$$\left(\frac{1-t}{1+t}\right)^{\frac{1}{2}\alpha} \frac{(1-t^2)\,dA_n^{(\alpha)}(t)/dt - (\alpha - nt)A_n^{(\alpha)}(t)}{(1-t^2)^{n+1}} \tag{7.4.6}$$

Multiplying by $(1-t^2)^2/t$ and using equation (6.3.17) reduces this expression to

$$\frac{\alpha^2 - n^2}{2n-1} \left(\frac{1-t}{1+t}\right)^{\frac{1}{2}\alpha} \frac{A_{n-1}^{(\alpha)}(t)}{(1-t^2)^{n-1}} \tag{7.4.7}$$

and gives

$$a_1 = -\frac{\alpha^2 - n^2}{2(2n-1)} \tag{7.4.8}$$

Since (7.4.7) is of the same form as (7.4.4) apart from a constant and replacing n by $n-1$, we may successively apply this process to give

$$a_r = \frac{(-1)^r \prod_{i=1}^{r} [\alpha^2 - (n+1-i)^2]}{2^r r! \prod_{i=1}^{r} [2n - 2i + 1]} \tag{7.4.9}$$

There are variations upon this solution dependent upon whether or not a factor $\sqrt{1-t^2}$ exists in the transfer function.[6.2] Again, even though finite-band solutions of this type may be obtained analytically, due to the non-optimum nature of the solution these will not be given.

In general, for any given specification, the phase and amplitude equalization techniques result in lower-degree transfer functions as compared to the exact linear phase case. However, the following sections provide solutions to the approximation problem which reduce the degree of the required filter by a significant amount.

7.5 OPTIMUM CONSTANT AMPLITUDE AND LINEAR PHASE FILTERS

Consider a transfer function of the form

$$S_{12}(t) = \frac{\sqrt{1-t^2}\, O_{2n-1}(t)}{D_{2n}(t)} \tag{7.5.1}$$

where $Q_{2n-1}(t)$ is an odd polynomial. This particular form of transfer function has been chosen due to its significance in being capable of being realized in the form of a generalized interdigital filter. This even-degree transfer function will initially be considered where the optimum number of maximally flat constraints are placed on the amplitude response and the remainder on the phase in a maximally flat manner around $t = \infty$.[7.4]

From the amplitude condition

$$\frac{1}{|S_{12}|^2} = 1 + \frac{K^2 \cos^n \omega}{\sin^2 \omega E^2 (\cos \omega)} \tag{7.5.2}$$

where $E(x)$ is an even polynomial.

Consider the factorization of (7.5.2) in the form

$$\frac{1}{S_{12}(t)S_{12}(-t)} = \left[1 + \frac{Kc^{2n}}{sE(c)}\right]\left[1 - \frac{Kc^{2n}}{sE(c)}\right] \tag{7.5.3}$$

where $c = \cosh p$, $s = \sinh p$ and writing

$$S_{12}(t) = \frac{\sqrt{1-t^2}\, O_{2n-1}(t)}{P_n(t, \sqrt{1-t^2})P_n(t, -\sqrt{1-t^2})} \tag{7.5.4}$$

we have

$$\frac{P_n(-t, \sqrt{1-t^2})P_n(t, -\sqrt{1-t^2})}{\sqrt{1-t^2}\, O_{2n-1}(t)} = 1 + \frac{Kc^{2n}}{sE(c)} \tag{7.5.5}$$

or

$$P_n(-t, \sqrt{1-t^2})P_n(t, -\sqrt{1-t^2}) = B[1 + \sqrt{1-t^2}\, O_{2n-1}(t)] \tag{7.5.6}$$

where $P_n(t, \sqrt{1-t^2})$ is an nth-degree polynomial in t and $\sqrt{1-t^2}$ and is always capable of being expressed as nth-degree in t and linear in $\sqrt{1-t^2}$.

Let $T_g(t, \sqrt{1+t^2})$ be the group delay associated with the polynomial $P_n(t, \sqrt{1-t^2})$ and if we require the group delay of S_{12} to be maximally flat,

$$T_g(t, \sqrt{1-t^2}) + T_g(t, -\sqrt{1-t^2})$$
$$= T_{g0}\left[1 - \frac{\cos^{2n}\omega E_1(\cos\omega)}{K^2\cos^{4n}\omega + \sin^2\omega E^2(\cos\omega)}\right] \quad (7.5.7)$$

using (7.5.4) and (7.5.2). From (7.5.6)

$$T_g(t, -\sqrt{1-t^2}) + T_g(-t, \sqrt{1-t^2})$$
$$= T_g(t, -\sqrt{1-t^2}) - T_g(t, \sqrt{1-t^2})$$
$$= (1-t^2)\text{Ev}\left\{\frac{[1+\sqrt{1-t^2}\,O_{2n-1}(t)]'}{[1+\sqrt{1-t^2}\,O_{2n-1}(t)]}\right\}$$
$$= \sqrt{1-t^2}\,\frac{(1-t^2)O'_{2n-1}(t) - tO_{2n-1}(t)}{1-(1-t^2)O^2_{2n-1}(t)}$$
$$= \frac{\cos^{2n-1}\omega E_2(\cos\omega)}{K^2\cos^{2n}\omega + \sin^2\omega E^2(\cos\omega)} \quad (7.5.8)$$

Subtracting (7.5.8) from (7.5.7) gives

$$T_g(t, \sqrt{1-t^2}) = \frac{T_{g0}}{2}\left[1 - \frac{\cos^{2n-1}\omega(a + b\cos\omega)}{E_3(\cos\omega)}\right] \quad (7.5.9)$$

Thus, the most general solution for $P_n(t, \sqrt{1-t^2})$ since it possesses a maximally flat group delay is

$$P_n(t, \sqrt{1-t^2}) = K_1 Q_n^{(\beta)}(t)(1 + at - b\sqrt{1-t^2})$$
$$+ K_1 Q_n^{(\beta+1)}(t)(1 - at + b\sqrt{1-t^2}) \quad (7.5.10)$$

where $Q_n^{(\beta)}(t)$ is the symmetrical Jacobi polynomial. Equation (7.5.10) is the combination of two linearly independent solutions to the maximally flat delay constraint, if the factor $(1 - at + b\sqrt{1-t^2})$ itself gives rise to a constant delay, and is of maximum degree n. The phase of the factor $(1 - at + b\sqrt{1-t^2})$ is

$$-\tan^{-1}\left(\frac{a\sin\omega}{b + \cos\omega}\right) \quad (7.5.11)$$

and is linear if and only if $b = 1$, $a = 1$ when it reduces to $\frac{1}{2}\omega$. Normalizing the leading coefficient to unity and the midband group delay to α gives

$$2\beta + 1 = \alpha, \quad K_1 = 1 \quad (7.5.12)$$

and hence

$$P_n(t, \sqrt{1-t^2}) = Q_n^{\frac{1}{2}(\alpha+1)}(t)(1 - t + \sqrt{1-t^2})$$
$$+ Q_n^{\frac{1}{2}(\alpha-1)}(t)(1 + t - \sqrt{1-t^2}) \qquad (7.5.13)$$

Finally, using (7.5.4) and (7.5.6) gives

$$S_{12}(t) = \frac{\mathrm{Od}[P_n(-t, \sqrt{1-t^2})P_n(t, -\sqrt{1-t^2})]}{P_n(t, \sqrt{1-t^2})P_n(t, -\sqrt{1-t^2})} \qquad (7.5.14)$$

and it may be shown that,[7.4]

$$\frac{1}{|S_{12}|^2} = 1 + \left\{\left[\frac{(\alpha^2-1)\alpha^2 a_{n-1} \cos^{2n} \omega}{2n \sin \omega E(\cos \omega)}\right]\right\}^2 \qquad (7.5.15)$$

where

$$E(\cos \omega) = \sum_{m=0}^{n-1} a_m \left(\frac{\cos \omega}{2}\right)^{2m} \qquad (7.5.16)$$

and

$$a_m = \frac{2a_{m-1}[\alpha^2 - (2n - 2m + 1)^2](n-m)}{m(2n-m)(2n-2m+1)} \qquad (7.5.17)$$

$$a_0 = 1$$

For the odd-degree case where

$$S_{12}(t) = \frac{O_{2n+1}(t)}{D_{2n+1}(t)} \qquad (7.5.18)$$

using similarly techniques to the lumped case, it may be shown that[7.4]

$$S_{12}(t) = \frac{\mathrm{Od}\{Q_n^{\frac{1}{2}\alpha}(-t)[2Q_{n+1}^{\frac{1}{2}\alpha}(t) - tQ_n^{\frac{1}{2}\alpha}(t)]\}}{Q_n^{\frac{1}{2}\alpha}(t)[2Q_{n+1}^{\frac{1}{2}\alpha}(t) - tQ_n^{\frac{1}{2}\alpha}(t)]} \qquad (7.5.19)$$

with the extra degree of freedom being placed upon the phase response and in both cases the low-pass solution is obtained by replacing t by $1/t$.

In addition to the various forms for band-pass or low-pass maximally flat response characteristics, similar solutions may be obtained for finite-band approximations. As an example we shall consider the construction of the low-pass odd-degree transfer function of the form

$$S_{12}(t) = \frac{E(t)}{P_1(t)P_2(t)} \qquad (7.5.20)$$

with

$$S_{12}(j \tan r\theta_0) = e^{2j\alpha r\theta_0} \qquad r = 0 \to n \qquad (7.5.21)$$

From the amplitude condition[7.5]

$$\frac{1}{S_{12}(t)S_{12}(-t)} = 1 - \left[\frac{t \prod_{r=1}^{n} (t^2 + \tan^2 r\theta_0)}{E(t)}\right]^2 \quad (7.5.22)$$

which from (7.5.20) gives

$$P_1(t)P_2(-t) = E(t) - t \prod_{r=1}^{n} (t^2 + \tan^2 r\theta_0) \quad (7.5.23)$$

From the phase condition

$$\text{Arg } P_1(j \tan r\theta_0) + \text{Arg } P_2(j \tan r\theta_0) = 2\alpha r\theta_0 \quad (7.5.24)$$

but from (7.5.23)

$$\text{Arg } P_1(j \tan r\theta_0) - \text{Arg } P_2(j \tan r\theta_0) = 0 \quad (7.5.25)$$

Thus, adding and subtracting (7.5.24) and (7.5.25) gives

$$\text{Arg } P_{1,2}(j \tan r\theta_0) = \alpha r\theta_0 \quad (7.5.26)$$

and consequently, in terms of the equidistant distributed linear phase polynomial,

$$P_1(t) = A_n^{(\alpha)}(t \mid \theta_0) \quad (7.5.27)$$

and

$$P_2(t) = (1 + K)A_{n+1}^{(\alpha)}(t \mid \theta_0) - KA_n^{(\alpha)}(t \mid \theta_0) \quad (7.5.28)$$

giving

$$S_{12}(t) = \frac{\text{Ev}[P_1(t)P_2(-t)]}{P_1(t)P_2(t)} \quad (7.5.29)$$

where K is a constant which may be chosen, as in Section 4.6, to provide an extra phase or amplitude condition.[7.5]

7.6 SYNTHESIS OF GENERALIZED INTERDIGITAL FILTERS

The interdigital filter described in Section 5.5 may be generalized by allowing coupling to exist between all of the wires. For this generalized interdigital filter,

$$S_{12}(t) = \frac{O(t)}{D(t)} \quad (7.6.1)$$

when the input and output are at the same end of the n-wire line and

$$S_{12}(t) = \frac{\sqrt{1-t^2}\, O(t)}{D(t)} \quad (7.6.2)$$

when they are at opposite ends. Realizability constraints and general synthesis processes have been developed in the general case.[7.1] However, transfer functions belonging to the class developed in Section 7.5 may be realized by symmetrical networks and this simplifies the synthesis process.[7.4] In this case,

$$S_{12}(t) = \frac{Y_e - Y_o}{(1 + Y_e)(1 + Y_o)} \qquad (7.6.3)$$

where Y_e and Y_o are the even- and odd-mode admittances formed by placing a magnetic and electric wall respectively along the plane of symmetry.

For the band-pass odd-degree transfer function providing an optimum constant amplitude and linear phase response,

$$S_{12}(t) = \frac{\mathrm{Od}[P_1(t)P_2(-t)]}{P_1(t)P_2(t)} \qquad (7.6.4)$$

and from (7.6.3) we may write

$$Y_o(t) = \frac{\mathrm{Ev}\, P_1(t)}{\mathrm{Od}\, P_1(t)} \qquad (7.6.5)$$

$$Y_e(t) = \frac{\mathrm{Ev}\, P_2(t)}{\mathrm{Od}\, P_2(t)} \qquad (7.6.6)$$

For a final realization in the form of a generalized interdigital filter, these reactance functions must be synthesized in the form of shunt short-circuited stubs separated by unit elements.[7.4] Due to the ability to scale the internal admittance level of this structure by transforming the stubs over the unit elements, the value of the u.e.s is not unique. It is convenient to assign a unity value to these u.e.s[7.4] and consequently the synthesis process is as follows.

Extract a shunt short-circuited stub from $Y_o(t)$ of characteristic admittance Y_{o1} to give a remaining admittance

$$Y'_o(t) = Y_o(t) - \frac{Y_{o1}}{t} \qquad (7.6.7)$$

Y_{o1} is chosen such that the subsequent extraction of a unit element will result in $Y'_o(1)$, the characteristic admittance of this unit element, having a unity value. Thus Y_{o1} is given by

$$Y_{o1} = Y_o(1) - 1 \qquad (7.6.8)$$

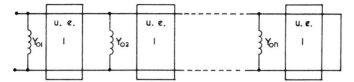

Figure 7.6.1 Odd-mode network for generalized interdigital filter

and after the extraction of the unit element

$$Y_o''(t) = \frac{Y_o'(t) - t}{1 - tY_o'(t)} \qquad (7.6.9)$$

which reduces the degree by one due to the cancellation of the factor $1 - t^2$. This entire cycle is repeated until the remaining admittance is infinite resulting in the overall odd-mode network shown in Figure 7.6.1. The same synthesis procedure is applied to $Y_e(t)$ which is one degree higher than $Y_o(t)$. However, when the remaining admittance is unity degree, realization is made in the form of a single shunt short-circuited stub as shown in Figure 7.6.2.

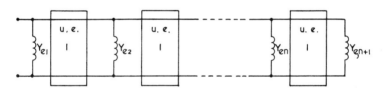

Figure 7.6.2 Even-mode network for generalized interdigital filter

The combination of the even- and odd-mode networks leads to an equivalent circuit for a generalized interdigital filter apart from $1 : -1$ transformers, as shown in Figure 7.6.3.

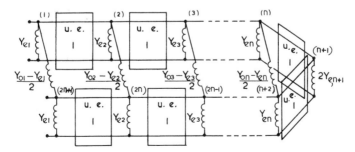

Figure 7.6.3 Equivalent circuit for generalized interdigital filter

The characteristic admittance matrix which describes this network is therefore

$$[Y] = \begin{bmatrix} 1 + \dfrac{Y_{e1} + Y_{o1}}{2} & -1 & 0 & 0 & 0 & \dfrac{-Y_{o1} + Y_{e1}}{2} \\ -1 & 2 + \dfrac{Y_{e2} + Y_{o2}}{2} & -1 & 0 & \dfrac{-Y_{o2} + Y_{e2}}{2} & 0 \\ 0 & -1 & 2 + \dfrac{Y_{e3} + Y_{o3}}{2} & -1 & \cdots & \cdots \\ 0 & 0 & -1 & \ddots & & \\ \vdots & \vdots & & & & \\ 0 & \dfrac{-Y_{o2} + Y_{e2}}{2} & & & 2 + \dfrac{Y_{e2} + Y_{o2}}{2} & -1 \\ \dfrac{-Y_{o1} + Y_{e1}}{2} & 0 & & & -1 & 1 + \dfrac{Y_{e1} + Y_{o1}}{2} \end{bmatrix} \quad (7.6.10)$$

which for realizability must be hyperdominant.[7.1] In this particular case, this condition reduces to

$$Y_{or} \geqslant Y_{er} \geqslant 0 \qquad r = 1 \to n + 1 \quad (7.6.11)$$

In general this does not necessarily hold. However, for the maximally

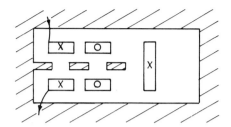

Figure 7.6.4 Physical realization of generalized interdigital filter ($n = 5$)

flat selective linear phase filters, this result has been found to be valid under the normal restriction for a bounded real function. For finite-band approximations, the result may not necessarily hold for large ripple levels. The final physical realization of this structure is illustrated in Figure 7.6.4 for a 5th-degree network.

There is however one practical disadvantage with this structure. At the open-circuited ends of the lines the fringing fields produce a reduction in the resonant frequency of the elements due to the effective capacitive loading. However, the resonant frequency of the coupling elements does not change significantly and therefore it becomes very difficult to construct the required commensurate network. This is one reason why the even-degree case is important, because resonant coupling elements may be entirely removed.[7.1]

For the band-pass even-degree transfer function providing an optimum constant amplitude and linear phase response,

$$S_{12}(t) = \frac{\mathrm{Od}[P(-t, \sqrt{1-t^2})P(t, -\sqrt{1-t^2})]}{P(t, \sqrt{1-t^2})P(t, -\sqrt{1-t^2})} \qquad (7.6.12)$$

and from (7.6.3) we may write

$$Y(t, \sqrt{1-t^2}) = \frac{P(t, \sqrt{1-t^2}) + P(-t, \sqrt{1-t^2})}{P(t, \sqrt{1-t^2}) - P(-t, \sqrt{1-t^2})} \qquad (7.6.13)$$

where

$$\begin{aligned} Y_e(t, \sqrt{1-t^2}) &= Y(t, \sqrt{1-t^2}) \\ Y_o(t, \sqrt{1-t^2}) &= Y(t, -\sqrt{1-t^2}) \end{aligned} \qquad (7.6.14)$$

These admittance functions may be synthesized in the form of a cascade of unit elements of unity characteristic admittance with shunt admittances at the junctions of each unit element.[7.4] These admittances consist of a single short-circuited shunt stub in parallel with an open- or short-circuited unit element in the even- and odd-mode cases respectively, which are half the physical length of the cascade unit elements. Therefore the general form for these shunt admittances, in the even-mode case, is

$$\frac{Y_L}{t} + Y_C \left(\frac{1 - \sqrt{1-t^2}}{t} \right) \qquad (7.6.15)$$

If $t' = \tanh \tfrac{1}{2}p$, i.e. $t = 2t'/(1 + t^2)$ then the second term reduces to $Y_C t'$.

From (7.6.13) let $Y(t, \sqrt{1-t^2})$ be expanded as

$$Y(t, \sqrt{1-t^2}) = \frac{E_1(t) + \sqrt{1-t^2}\, E_2(t)}{O_2(t) + \sqrt{1-t^2}\, O_2(t)} \qquad (7.6.16)$$

and the extraction of the first shunt admittance yields

$$Y'(t, \sqrt{1-t^2}) = Y(t, \sqrt{1-t^2}) - \frac{Y_{L1}}{t} - Y_{C1} \frac{1-\sqrt{1-t^2}}{t} \qquad (7.6.17)$$

For the adjacent unit element to possess a unity characteristic admittance we must have

$$1 + Y_{L1} - Y_{C1} = Y(1,0) \qquad (7.6.18)$$

Furthermore, in order to obtain a reduction in degree upon extraction of this unit element, we require that the transfer admittance of the overall network given by

$$Y_{12}' = \frac{Y'(t, -\sqrt{1-t^2}) - Y'(t, \sqrt{1-t^2})}{2}$$

$$= \frac{Y(t, -\sqrt{1-t^2}) - Y(t, \sqrt{1-t^2})}{2} - \frac{Y_{C1}\sqrt{1-t^2}}{t} \qquad (7.6.19)$$

possesses a factor $(1-t^2)^{3/2}$. Thus,

$$Y_{C1} = \left. \frac{Y(t, -\sqrt{1-t^2}) - Y(t, \sqrt{1-t^2})}{2\sqrt{1-t^2}} \right|_{t=1} \qquad (7.6.20)$$

The extraction of this adjacent unit element therefore gives a remaining admittance

$$Y''(t, \sqrt{1-t^2})$$
$$= \frac{Y(t, \sqrt{1-t^2}) - [(Y_{L1} + Y_{C1})/t] + Y_{C1}[\sqrt{1-t^2}/t] - t}{1 - tY(t, \sqrt{1-t^2}) + (Y_{L1} + Y_{C1}) - Y_{C1}\sqrt{1-t^2}}$$

$$(7.6.21)$$

The rational part of the numerator of $Y''(t, \sqrt{1+t^2})$ is given by the zeros of

$$\frac{Y(t,\sqrt{1-t^2}) + Y(t,-\sqrt{1-t^2})}{2} - \frac{Y_{L1} + Y_{C1}}{t} - t \qquad (7.6.22)$$

which vanishes at $t = \pm 1$ due to (7.6.18). The irrational part of this numerator is

$$\sqrt{1-t^2} \left[\frac{Y(t,\sqrt{1-t^2}) - Y(t,-\sqrt{1-t^2})}{2\sqrt{1-t^2}} + \frac{Y_{C1}}{t} \right] \qquad (7.6.23)$$

and from (7.6.20) contains the factor $(1-t^2)^{3/2}$.

Since a similar result holds for the denominator of $Y''(t,\sqrt{1-t^2})$, a factor $1-t^2$ cancels between the numerator and denominator thus leaving $Y''(t,\sqrt{1-t^2})$ in the same form as $Y(t,\sqrt{1-t^2})$ but of one degree less. This cycle in the synthesis procedure may be repeated until the remaining admittance is of unity degree and of the form

$$\frac{Y_{Ln}}{t} + Y_{Cn} \frac{1-\sqrt{1-t^2}}{t} \qquad (7.6.24)$$

which may be identified as a typical shunt susceptance.

Utilizing the relationship between the even- and odd-mode admittance functions, the overall network as shown in Figure 7.6.5. may be recovered. This may be identified, apart from 1:−1 transformers as the equivalent circuit of a generalized interdigital filter[7.1] with a characteristic admittance matrix

$$\begin{bmatrix} Y_{L1} + Y_{C1} + 1 & -1 & 0 & 0 & 0 & -Y_{C1} \\ -1 & Y_{L2} + Y_{C2} + 2 & -1 & 0 & -Y_{C2} & 0 \\ 0 & -1 & Y_{L3} + Y_{C3} + 2 & -1 & | & | \\ | & | & | & | & | & | \\ 0 & -Y_{C2} & | & | & | & | \\ -Y_{C1} & 0 & | & Y_{L2} + Y_{C2} + 2 & -1 \\ & & & -1 & Y_{L1} + Y_{C1} + 1 \end{bmatrix} \qquad (7.6.25)$$

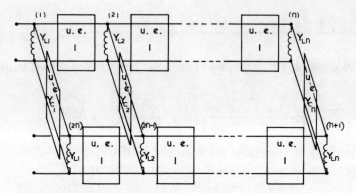

Figure 7.6.5 Equivalent circuit for even-degree generalized interdigital filter

and realizability requires that

$$Y_{Lr} \geq 0, \qquad Y_{Cr} \geq 0 \qquad r = 0 \to n \qquad (7.6.26)$$

Again, as in the odd-degree case, this does not necessarily hold but for most selective linear phase transfer functions the network is normally realizable. The final physical structure is shown in Figure 7.6.6 for a 6th-degree network and does not suffer from the fringing field effects other than in an overall change in resonant frequency.

To illustrate the design process, we shall synthesize a 6th-degree filter which possesses a transfer function with an optimum maximally flat amplitude characteristic and linear phase response. From equation (7.5.13)

$$P(t, \sqrt{1-t^2}) = Q_3^{\frac{1}{2}(\alpha+1)}(t)(1 - t + \sqrt{1-t^2}) \\ + Q_3^{\frac{1}{2}(\alpha-1)}(t)(1 + t - \sqrt{1-t^2}) \qquad (7.6.27)$$

and $S_{12}(t)$ is given by (7.6.12).

Initially we must generate $Q_3^{(\beta)}(t)$ using the recurrence formula deduced from (6.3.10) and (6.3.29), i.e.

$$Q_{n+1}^{(\beta)}(t) = tQ_n^{(\beta)}(t) + \frac{\beta^2 - n^2}{4n^2 - 1} Q_{n-1}^{(\beta)}(t) \qquad (7.6.28)$$

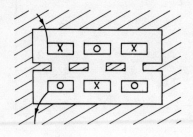

Figure 7.6.6 Physical realization of generalized interdigital filter ($n = 6$)

with the initial conditions
$$Q_0^{(\beta)}(t) = 1, \qquad Q_1^{(\beta)}(t) = t + \beta \qquad (7.6.29)$$
Thus,
$$Q_2^{(\beta)}(t) = t(t + \beta) + \frac{\beta^2 - 1}{3} \cdot 1$$
$$= t^2 + \beta t + \frac{\beta^2 - 1}{3} \qquad (7.6.30)$$
and
$$Q_3^{(\beta)}(t) = t\left(t^2 + \beta t + \frac{\beta^2 - 1}{3}\right) + \frac{\beta^2 - 4}{15}(t + \beta)$$
$$= t^3 + \beta t^2 + \frac{2\beta^2 - 3}{5} t + \frac{\beta(\beta^2 - 4)}{15} \qquad (7.6.31)$$

Substitution into (7.6.27) yields
$$P(t, \sqrt{1-t^2}) = 2\left[t^3 + \frac{3\alpha}{5} t^2 + 3\frac{\alpha^2 - 5}{20} t + \frac{\alpha(\alpha^2 - 13)}{60}\right.$$
$$\left. + \sqrt{1-t^2}\left(t^2 + \frac{2\alpha}{5} t + \frac{\alpha^2 - 5}{20}\right)\right] \qquad (7.6.32)$$

which from (7.6.13) gives
$$Y(t, \sqrt{1-t^2}) =$$
$$\frac{(3\alpha/5)t^2 + [\alpha(\alpha^2 - 13)/60] + \sqrt{1+t^2}\,[t^2 + (\alpha^2 - 5)/20]}{t^3 + 3[(\alpha^2 - 5)/20]t + (2\alpha/5)t\sqrt{1-t^2}}$$
$$(7.6.33)$$

Extraction of the typical shunt admittance given by (7.6.17) yields
$$Y'(t, \sqrt{1-t^2}) = Y(t, \sqrt{1-t^2}) - \frac{Y_{L1} + Y_{C1}}{t} + \frac{Y_{C1}\sqrt{1-t^2}}{t}$$
$$(7.6.34)$$

where from (7.6.18), (7.6.20) and (7.6.33)
$$1 + Y_{L1} + Y_{C1} = \frac{\alpha(\alpha^2 + 23)}{3(3\alpha^2 + 5)} \qquad (7.6.35)$$

and
$$Y_{C1} = \frac{(\alpha^2 - 25)(\alpha^2 - 9)}{3(3\alpha^2 + 5)^2} \qquad (7.6.36)$$

Extracting a unit element of unity characteristic admittance gives

$$Y''(t,\sqrt{1-t^2}) = \frac{Y(t,\sqrt{1-t^2}) - [(\alpha-1)(\alpha-3)(\alpha-5)/(3(3\alpha^2+5))t] + [(\alpha^2-25)(\alpha^2-9)/(3(3\alpha^2+5)^2)](\sqrt{1-t^2}/t)}{1 - tY(t,\sqrt{1-t^2}) + [(\alpha-1)(\alpha-3)(\alpha-5)/(3(3\alpha^2+5))] - [(\alpha^2-25)(\alpha^2-9)/(3(3\alpha^2+5)^2)]\sqrt{1-t^2}}$$
(7.6.37)

After cancellation of the common factor $1-t^2$ between numerator and denominator, this reduces to

$$Y''(t,\sqrt{1-t^2}) = \frac{E_1 + \sqrt{1-t^2}\,E_2}{O_1 + \sqrt{1-t^2}\,O_2} \quad (7.6.38)$$

where

$$E_1 = t^2 + \frac{\alpha(\alpha^2-13)}{60} - \frac{(\alpha^2-5)(\alpha-1)(\alpha-3)(\alpha-5)}{20(3\alpha^2+5)}$$

$$\qquad - \frac{2\alpha(\alpha^2-25)(\alpha^2-9)}{15(3\alpha^2+5)^2}$$

$$E_2 = \frac{2\alpha}{5} + \frac{(\alpha^2-25)(\alpha^2-9)}{3(3\alpha^2+5)^2} - 1$$

$$O_1 = \left[\frac{3\alpha}{5} - \frac{(\alpha-1)(\alpha-3)(\alpha-5)}{3(3\alpha^2+5)}\right]t \qquad (7.6.39)$$

$$O_2 = \left[1 - \frac{(\alpha^2-25)(\alpha^2-9)}{3(3\alpha^2+5)^2}\right]t$$

Reapplying the synthesis cycle gives

$$Y_{L2} + Y_{C2} = \frac{3(3\alpha^2+5)^3}{16\alpha(\alpha^2-1)(17\alpha^2-25)} - 2 \qquad (7.6.40)$$

and

$$Y_{C2} = \frac{3(\alpha^2-25)(3\alpha^2+5)(77\alpha^2+75)}{32\alpha^2(17\alpha^2-25)^2(\alpha^2-1)} \qquad (7.6.41)$$

and leaves the shunt admittance

$$\frac{Y_{L3}+Y_{C3}}{2} - \frac{Y_{C3}\sqrt{1-t^2}}{t} \qquad (7.6.42)$$

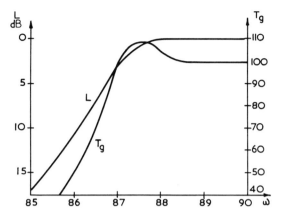

Figure 7.6.7 Computed response of maximally flat amplitude and phase generalized interdigital filter ($n = 6$)

where

$$Y_{L3} + Y_{C3} = \frac{64\alpha^3(\alpha^2 - 1)(17\alpha^2 - 25)^3}{15.25.25(3\alpha^2 + 5)^4(\alpha^2 - 9)} - 1 \qquad (7.6.43)$$

and

$$Y_{C3} = \frac{32\alpha^2(\alpha^2 - 1)(13\alpha^2 + 75)(17\alpha^2 - 25)^2}{3.25.25(3\alpha^2 + 5)^4(\alpha^2 - 9)} \qquad (7.6.44)$$

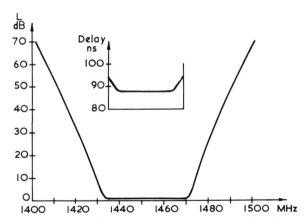

Figure 7.6.8 Measured response of maximally flat generalized interdigital filter ($n = 14$ plus transformer elements)

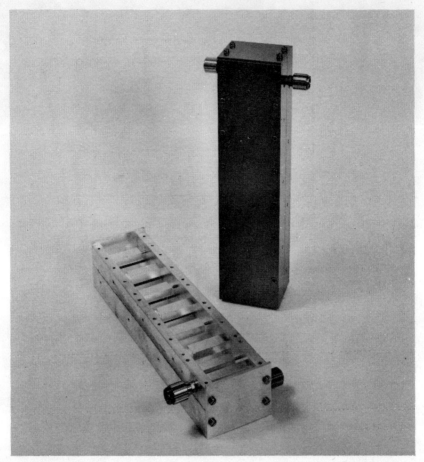

Figure 7.6.9   Physical construction of generalized interdigital filter

from which the characteristic admittance matrix of the generalized interdigital network may be directly formed. Inspection of the element values reveals that the network is theoretically realizable if $\alpha \geqslant 5$. For large values of α when narrow bandwidth responses are required, redundant transformer elements must be introduced, as in the conventional interdigital filter, to provide a physically realizable device.

In Figure 7.6.7, the response characteristic for this filter with $\alpha = 100$ is shown and in Figure 7.6.8 the measured response characteristic of a typical practical filter designed on a maximally flat basis is shown. This 14th-degree filter, with redundant transformer elements is illustrated in Figure 7.6.9. Finally, in Figure 7.6.10 the measured characteristics of a 14th-degree generalized interdigital filter designed on a finite-band approximation for constant amplitude and interpolation

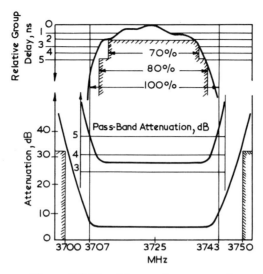

Figure 7.6.10 Measured response of generalized interdigital filter designed to interpolate to an ideal amplitude and arbitrary prescribed phase

to a specified phase performance associated with a small quadratic group delay distortion is shown, illustrating the type of performance which can be achieved in practice.

CHAPTER 8

Digital Filters

8.1 INTRODUCTION

The transfer function for a digital filter may be written as $S_{12}(z)$ where $z = e^{-Tp}$ with T the sampling period of the system. Although $S_{12}(z)$ must be stable and therefore devoid of poles in $|z| \leqslant 1$, a digital filter is naturally non-reciprocal and can readily be active. Thus, apart from any advantage which might be exploited with regard to the non-reciprocal nature of the devices, the solution to the approximation problem is identical to that for distributed networks. The essential difference is in the realization and, since the devices are potentially active, the problem of transfer-function sensitivity with respect to individual elements in the realization arises.

Using multipliers, adders and basic delay elements a transfer function

$$S_{12}(z) = \frac{\sum\limits_{r=0}^{n} a_r z^r}{1 + \sum\limits_{r=1}^{n} b_t z^r} = \frac{A(z)}{B(z)} \qquad (8.1.1)$$

has the direct realization shown in Figure 8.1.1.

If $B(z) = 1$, then we have a non-recursive filter (finite duration impulse response) which is inherently stable and capable of providing an exact linear phase response at all frequencies (see Section 7.2). As distinct from the distributed case, the realization is very simple and therefore this class of transfer functions is important even though high-degree filters are required for reasonably selective filters and are more sensitive with regard to element (multiplier) values than any distributed realization. If $B(z) \neq 1$, then the resulting recursive filter becomes even more sensitive and due to non-linear rounding-off effects in multiplication is potentially unstable in very selective response cases.

An improvement is obtained if the transfer function is factored into quadratic sections which are then realized individually. However, the sensitivity is still greater than any equivalent distributed realization and the question arises as to whether realizations exist which imitate the sensitivity characteristics of passive distributed filters. Such filters do exist and have been termed wave digital or digital wave filters.[8.1] In the

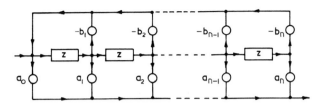

Figure 8.1.1   Direct realization of a digital filter

following section, details are provided for the design belonging to this class which is based upon the basic distributed prototype filter consisting of a cascade of unit elements.

In the design of selective linear phase filters for distributed filters, the transfer functions were constrained to be realizable by passive devices with a direct reciprocal realization. In Section 8.3 it is shown that a reduction in degree for a given specification can be achieved if this restriction is removed. The maximally flat and equidistant approximations to constant amplitude and linear phase responses are developed with this additional flexibility.

8.2 BASIC DIGITAL WAVE FILTERS

The distributed low-pass prototype filter with a Chebyshev response characteristic is described by equation (5.2.4) where

$$|S_{12}|^2 = \frac{1}{1 + \epsilon^2 T_n^2(\sin \omega/\alpha)} \quad (8.2.1)$$

and ω has been normalized. The realization may be made in the form of a cascade of unit elements of unity characteristic admittance with ideal transformers at the junctions and, from Section 5.3, explicit formulas deduced. The scattering matrix for the rth line normalized to unity terminations is

$$\begin{bmatrix} 0 & z^{1/2} \\ z^{1/2} & 0 \end{bmatrix} \quad (8.2.2)$$

and for the rth transformer, between the rth and $(r+1)$th lines,

$$\begin{bmatrix} \gamma_r & \sqrt{1-\gamma_r^2} \\ \sqrt{1-\gamma_r^2} & -\gamma_r \end{bmatrix} \quad (8.2.3)$$

where

$$\gamma_r = \frac{n_r^2 - 1}{n_r^2 + 1} \quad (8.2.4)$$

are the junction reflection coefficients.

From Figure 5.2.2 for a filter of degree $2n$ or $2n+1$, at a finite number of points ω_r in the passband

$$|S_{12}|^2_{\omega=\omega_r} = 1 \qquad r = 1 \to n \qquad (8.2.5)$$

and since the device is passive $|S_{12}|^2 \leqslant 1$. Now if all of the junction reflection coefficients are realized exactly, we obtain the exact response given in Figure 5.2.2. If, however, one of the junction reflection coefficients γ_i possesses a value slightly smaller than the required value, then

$$|S_{12}|^2_{\omega=\omega_r} \leqslant 1 \qquad r = 1 \to n \qquad (8.2.6)$$

since $|S_{12}|^2 \leqslant 1$. Similarly, if γ_i is slightly larger then the same result (8.2.6) must hold. Thus

$$\left. \frac{\partial |S_{12}|^2}{\partial \gamma_i} \right|_{\omega=\omega_r} = 0 \qquad r = 1 \to n,\ i = 0 \to n \qquad (8.2.7)$$

that is, the first-order sensitivities of the transfer function with respect to the junction reflection coefficients are zero at n finite points. Thus, if a digital filter can be constructed which models this distributed filter, then the sensitivity with respect to the multiplier values will be low.

The reason for the low sensitivity may be explained if the matrix (8.2.3) is examined closely. The passive, lossless nature of this section results from the fact that the transfer and reflection parameters are uniquely related by a quadratic equation independent of the value of γ_r. If a direct digital realization is modelled on this distributed filter, then the numbers γ_r and $\sqrt{1-\gamma_r^2}$ would have to be realized by separate multipliers and hence the sensitivity properties would disappear immediately. However, a non-reciprocal scattering matrix of the form

$$\begin{bmatrix} \gamma_r & 1+\gamma_r \\ 1-\gamma_r & -\gamma_r \end{bmatrix} \qquad (8.2.8)$$

may be realized directly by one multiplier as illustrated in Figure 8.2.1.

For any signal which passes in both directions through the sections described by (8.2.3) and (8.2.8) the double transfer function is the same $1-\gamma_r^2$. For signals reflected directly from either side of these sections the reflection coefficient is the same. Thus, the overall

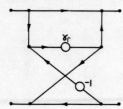

Figure 8.2.1 Digital realization of scattering matrix (8.2.8)

reflection coefficient $S_{11}(z)$ for the network using sections described by (8.2.3) is the same as that produced by the network composed of sections described by (8.2.8). The transfer function $S_{12}(z)$ must be modified, however, to

$$S'_{12}(z) = S_{12}(z) \prod_{r=0}^{n} \left(\frac{1+\gamma_r}{1-\gamma_r}\right)^{1/2} \qquad (8.2.9)$$

and therefore the sensitivity properties of (8.2.8) are only retained within a constant gain multiplier. However, since only the gain variation with respect to frequency is important, the essential sensitivity properties have been retained.

In order to reduce the number of delay elements, the matrix (8.2.2) is also replaced by the non-reciprocal matrix with one delay element in the form

$$\begin{bmatrix} 0 & z \\ 1 & 0 \end{bmatrix} \qquad (8.2.10)$$

which retains the same overall reflection coefficient but the transfer function is modified to

$$S''_{12}(z) = S_{12}(z) z^{\frac{1}{2}n} \prod_{r=0}^{n} \left(\frac{1+\gamma_r}{1-\gamma_r}\right)^{1/2} \qquad (8.2.11)$$

which only produces an extra constant delay.

Thus, a digital wave filter may be designed based upon the basic distributed prototype filter with a circuit configuration shown in Figure 8.2.2 where the multiplier values are all less than unity and in terms of the element values of the prototype

$$\gamma_r = (-1)^r \frac{Y_r Y_{r+1} - K_{r,r+1}^2}{Y_r Y_{r+1} + K_{r,r+1}}$$

$r = 0 \to n$ with $Y_o = K_{0,1} = 1$, and Y_r and $K_{r,r+1}$ are given by equations (5.4.5) and (5.4.6) for the Chebyshev filter.

More sophisticated digital wave filters may be constructed from other distributed prototypes, but in order to retain the basic sensitivity properties it is essential that the direct reflection coefficients and the two-way transmission coefficients be modelled section by section using

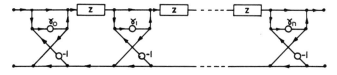

Figure 8.2.2   Basic digital wave filter realization

a canonic number of multipliers. If this is not the case, then the low-sensitivity properties associated with passive networks will not be retained.

8.3 SELECTIVE LINEAR PHASE FILTERS

The design of digital selective linear phase filters is essentially the same as the design of distributed filters. However, with digital filters it is normally simpler to realize non-reciprocal transfer functions and gains greater than unity are possible with these active filters. However, the stability problem is the same. In terms of the variable $t = \tanh p$, p normalized to $\tfrac{1}{2}T$, we may therefore have transfer functions of the form

$$S_{12}(t) = \frac{N(t)}{D_n(t)} \tag{8.3.1}$$

where $N(t)$ is an arbitrary polynomial and $D_n(t)$ is constrained to be an nth-degree Hurwitz polynomial

Consider the low-pass transfer function defined by[7.5]

$$S_{12}(\infty) = 0 \tag{8.3.2}$$

and

$$S_{12}(j \tan r\theta_0) = e^{-2j\alpha_r\theta_0} \qquad r = 0 \to n-1 \tag{8.3.3}$$

That is an equidistant interpolation to unity amplitude and a linear phase response. From (8.3.2) $N(t)$ must be of degree $n-1$ and from (8.3.3)

$$N(t)N(-t) - D_n(t)D_n(-t) = 2Kt^2 \prod_{r=1}^{n-1} (t^2 + \tan^2 r\theta_0) \tag{8.3.4}$$

and

$$\operatorname{Arg} D_n(j \tan r\theta_0) - \operatorname{Arg} N(j \tan r\theta_0) = 2\alpha r\theta_0 \qquad r = 0 \to n-1 \tag{8.3.5}$$

Let $D_n(t)$ be expressed as

$$D_n(t) = p(t) + Ktq(t) \tag{8.3.6}$$

where

$$q(t)p(-t) - q(-t)p(t) = t \prod_{r=1}^{n-1} (t^2 + \tan^2 r\theta_0) \tag{8.3.7}$$

then from (8.3.4)

$$N(t)N(-t) = [p(t) + Ktq(t)][p(-t) - Ktq(-t)]$$
$$- 2Kt \prod_{r=1}^{n-1} (t^2 + \tan^2 r\theta_0)$$
$$= [p(t) - Ktq(t)][p(-t) + Ktq(-t)] \quad (8.3.8)$$

and we may identify

$$N(t) = p(-t) + Ktq(-t) \quad (8.3.9)$$

Thus, from (8.3.6) and (8.3.9), $D_n(t)$ and $N(t)$ will satisfy (8.3.5) if

$$\text{Arg } p(j \tan r\theta_0) = \text{Arg } q(j \tan r\theta_0) = \alpha r\theta_0 \quad r = 0 \to n-1$$
$$(8.3.10)$$

Thus, if $p(t) = A_n^{(\alpha)}(t \mid \theta_0)$ and $q(t) = A_{n-1}^{(\alpha)}(t \mid \theta_0)$, the equidistant distributed linear phase polynomials, then (8.3.10) is satisfied together with (8.3.7) due to the recurrence formula (6.4.9). Hence,

$$S_{12}(t) = \frac{A_n^{(\alpha)}(-t \mid \theta_0) + KtA_{n-1}^{(\alpha)}(-t \mid \theta_0)}{A_n^{(\alpha)}(t \mid \theta_0) + KtA_{n-1}^{(\alpha)}(t \mid \theta_0)} \quad (8.3.11)$$

where

$$K = \left. \frac{A_n^{(\alpha)}(t \mid \theta_0)}{tA_{n-1}^{(\alpha)}(t \mid \theta_0)} \right|_{t=\infty} \quad (8.3.12)$$

in order to satisfy (8.3.2) and is stable when $A_n^{(\alpha)}(t \mid \theta_0)$ is Hurwitz.*
For $\theta_0 = 0$ we recover the maximally flat solution[8.2] where the first $2n - 1$ derivatives on amplitude and phase deviation from linearity vanish at $t = 0$.

A similar construction may be used to interpolate in an optimum manner to any prescribed characteristic with this class of recursive response characteristics providing a significant reduction in degree over the non-recursive case for a selective linear phase characteristic.[7.5]

---

*N.B. $|S_{12}|^2 \not\leqslant 1$ since $1 - |S_{12}|^2$ changes sign when the frequency moves through a point of interpolation.

APPENDIX

Miscellaneous Amplitude Approximations

A.1 GENERALIZED CHEBYSHEV FUNCTIONS WITH PRESCRIBED POLES

In some restricted classes of networks where the class is defined by the location of the transmission zeros, an equiripple passband amplitude characteristic may be required. In this case,

$$|S_{12}(j\omega)|^2 = \frac{A}{1 + \epsilon^2 C_n^2(\omega)} \tag{A.1.1}$$

where $C_n(x)$ is a function of the form

$$C_n(x) = \frac{P_n(x)}{E(x)} \tag{A.1.2}$$

with $P_n(x)$ an nth-degree polynomial and $E(x)$ an even polynomial of $2m$ less than or equal to n which is defined by the location of the transmission zeros. The problem is then to determine $C_n(x)$ which is equiripple in the interval $|x| \leq 1$ between the limits ± 1 with the maximum number of turning points in the interval $|x| \leq 1$.[A.1] Thus it follows that $C_n(x)$ must satisfy a differential equation of the form

$$\frac{dC_n(x)}{dx} = \frac{\sqrt{C_n^2(x) - 1}}{\sqrt{x^2 - 1}} \frac{Q(x)}{E(x)} \tag{A.1.3}$$

since the denominator of $dC_n(x)/dx$ must contain the factor $E^2(x)$ and $Q(x)$ is uniquely determined from the condition that $C_n(x)$ is a rational function of degree n.

Let

$$C_n(x) = \cosh U \tag{A.1.4}$$

and

$$x = \frac{1}{\sqrt{1 - z^2}} \tag{A.1.5}$$

or

$$z = \sqrt{1 - \frac{1}{x^2}} \tag{A.1.6}$$

Thus,

$$\frac{dC_n}{dU} = -\frac{1}{x^2\sqrt{x^2-1}} = -\frac{(z^2-1)}{z} \tag{A.1.7}$$

Substitution into equation (A.1.3) gives

$$\frac{dU}{dz} = \frac{B(z)}{(1-z^2)E_1(z)} \tag{A.1.8}$$

where

$$E_1(z) = E([1-z^2]^{-1/2})(1-z^2)^m$$

$$= K \prod_{r=1}^{m} (z_i^2 - z^2) \tag{A.1.9}$$

thus reducing (A.1.8) to

$$\frac{dU}{dz} = \sum_{r=0}^{m} \left(\frac{k_i}{z_i - z} + \frac{k_i'}{z_i + z} \right) + K(z) \tag{A.1.10}$$

with $z_0 = 1$ and $K(z)$ a polynomial in z under the assumption that the z_i are distinct. Hence

$$C_n(x) = \cosh\left\{ \sum_{r=0}^{m} [k_i' \ln(z_i + z) - k_i \ln(z_i - z)] + \int K(z) \right\} \tag{A.1.11}$$

with z given by (A.1.6). For $C_n(x)$ to be a rational function in x then $k_i = k_i' = 1$, $k_0 = 0$ or $\tfrac{1}{2}$ and $K(z) = 0$. Hence

$$C_n(x) = \frac{1}{2}\left[\frac{H(-z)}{H(z)}\left(\frac{1-z}{1+z}\right)^{\tfrac{1}{2}q} + \frac{H(z)}{H(-z)}\left(\frac{1+z}{1-z}\right)^{\tfrac{1}{2}q} \right] \tag{A.1.12}$$

where

$$H(z) = \prod_{r=1}^{m} (z_i + z) \tag{A.1.13}$$

and $q = 0$ or 1. Additionally, we also require that there are the maximum number of turning points in the interval $|x| \leq 1$; that is along the entire imaginary axis of z. Let $z = j\Omega$, then

$$C_n = \cos\theta$$

$$\theta = 2\,\mathrm{Arg}[H(j\Omega)(1+j\Omega)^{\tfrac{1}{2}q}] \tag{A.1.14}$$

and as Ω moves from $-\infty \to +\infty$ we require the maximum variation in θ which implies that $H(z)$ must be a strict Hurwitz polynomial in z. Thus, in summary, we may write

$$C_n(x) = \frac{1}{2}\left[\frac{M(-z)}{M(z)} + \frac{M(z)}{M(-z)}\right]$$

$$= \frac{P_n(x)}{E(x)} \qquad (A.1.15)$$

with

$$z = \sqrt{1 - \frac{1}{x^2}} \qquad (A.1.16)$$

and $M(z)$ is strictly Hurwitz and defined by

$$M(z)M(-z) = E([1-z^2]^{1/2})(1-z^2)^{\frac{1}{2}n} \qquad (A.1.17)$$

A.2 GENERALIZED CHEBYSHEV FUNCTIONS WITH PRESCRIBED ZEROS

Consider the even-degree function $F_{2n}(x)$ which is optimally equiripple between the limits ± 1 for $|x| \leq 1$ and

$$\begin{aligned}F_{2n}(\pm x_i) &= 0 \\ 0 < x_i &< x_{i+1} < 1\end{aligned} \quad i = 1 \to n \qquad (A.2.1)$$

This problem could arise when passband characteristics are approximated by interpolation and an optimum equiripple amplitude characteristic is required. Unfortunately in general, the zeros of $C_n(x)$ given by (A.1.15) may not be obtained in an analytical manner and therefore direct use of the result derived in Section A.1 cannot be used. However, we shall prove that we may obtain $F_{2n}(x)$ analytically from,[A.2]

$$\frac{1}{F_{2n}(x)} = \frac{1}{2}\left[\frac{R(-\lambda)}{R(\lambda)} + \frac{R(\lambda)}{R(-\lambda)}\right] \qquad (A.2.2)$$

where

$$\lambda = \sqrt{\frac{1}{x^2} - 1} \qquad (A.2.3)$$

and

$$R(\lambda) = \prod_{r=1}^{n}[\lambda + (-1)^r \lambda_r] \qquad (A.2.4)$$

with

$$\lambda_r = +\sqrt{\frac{1}{x_r^2} - 1} \qquad r = 1 \to n \qquad (A.2.5)$$

and

$$0 < x_r < x_{r+1} < 1 \qquad (A.2.6)$$

Let $\lambda = jz$ and

$$R(jz) = \frac{(1-j)}{\sqrt{2}} [M(z) + jM(-z)] \qquad (A.2.7)$$

Hence,

$$F_{2n}(x) = \frac{2R(\lambda)R(-\lambda)}{R^2(\lambda) + R^2(-\lambda)}$$

$$= \frac{1}{2} \left[\frac{M^2(-z) + M^2(z)}{M(z)M(-z)} \right]$$

$$= \frac{1}{2} \left[\frac{M(-z)}{M(z)} + \frac{M(z)}{M(-z)} \right] \qquad (A.2.8)$$

Thus, if $M(z)$ is an nth-degree Hurwitz polynomial, then $F_{2n}(x)$ is optimally equiripple in the interval $|x| \leq 1$. From equation (A.2.7), by taking the conjugate, we have

$$R(-jz) = \frac{1+j}{\sqrt{2}} [M(z) - jM(-z)] \qquad (A.2.9)$$

and therefore

$$M(z) = \frac{R(jz) - jR(-jz)}{\sqrt{2}(1-j)} \qquad (A.2.10)$$

Taking the conjugate of this expression gives

$$M^*(z) = \frac{R(-jz) + jR(jz)}{\sqrt{2}(1+j)}$$

$$= \frac{R(jz) - jR(-jz)}{\sqrt{2}(1-j)} \qquad (A.2.11)$$

revealing that $M(z)$ is a polynomial with real coefficients. Furthermore,

$$Z(z) = \frac{-jR(-jz)}{R(jz)} = -j \prod_{r=1}^{n} \left[\frac{-jz + (-1)^r \lambda_r}{jz + (-1)^r \lambda_r} \right]$$

$$= (-1)^{n-1} j + \sum_{r=1}^{n} \frac{k_r}{z - j(-1)^r \lambda_r} \qquad (A.2.12)$$

and $k_r > 0$ since the λ_r are distint and interlaced. Hence, $Z(z)$ is a positive function and the zeros of $1 + Z(z)$ must lie in the strict left-half plane, resulting in $M(z)$ being a strict Hurwitz polynomial.

A.3 EVEN POLYNOMIALS AND FUNCTIONS EQUIRIPPLE OVER TWO BANDS

Consider the problem of obtaining the even polynomial or function $F_{2n}(x)$ which is equiripple between ± 1 in the two bands $-1 < x < -x_0$, $x_0 < x < 1$. In the polynomial case we immediately obtain

$$F_{2n}(x) = T_{2n}\left(\frac{\sqrt{x^2 - x_0^2}}{\sqrt{1 - x_0^2}}\right) \tag{A.3.1}$$

as the required even polynomial of degree $2n$ where $T_{2n}(x)$ is the Chebyshev polynomial. A similar transformation of argument also applies in the generalized rational Chebyshev function case. If a constant level is required to be approximated, then the appropriate constant may be added to (A.3.1) to produce the required result. Thus, in the even-function case, the transformation to the two-band approximation of this type is trivial.

However, in many network problems it may be desirable to retain a transmission zero or perfect transmission at the origin. In most problems this reduces to obtaining the odd polynomial or function which approximates to a constant level over a finite band. The solutions to this problem are treated in the remaining sections.

A.4 MAXIMALLY FLAT ODD POLYNOMIAL APPROXIMATING A CONSTANT

Consider the odd polynomial $P_{2n+1}(x)$ of degree $2n + 1$ which is required to approximate to unity in a maximally flat manner around $x = 1$. Then,

$$\frac{d P_{2n+1}(x)}{dx} = K(1 - x^2)^n \tag{A.4.1}$$

since $P_{2n+1}(x)$ is an even polynomial of degree $2n$. Integrating directly yields,

$$P_{2n+1}(x) = \frac{\int_0^x (1 - y^2)^n \, dy}{\int_0^1 (1 - y^2)^n \, dy} \tag{A.4.2}$$

since $P_{2n+1}(1) = 1$.

Let

$$I_n(x) = \int_0^x (1-y^2)^n \, dy \qquad (A.4.3)$$

Integration by parts gives

$$I_n(x) = \frac{2n}{2n+1} I_{n-1}(x) + (1-x^2)^n x \qquad (A.4.4)$$

and therefore

$$I_n(1) = \frac{2^{2n}(n!)^2}{(2n)!} \qquad (A.4.5)$$

Thus, from (A.4.2), (A.4.4) and (A.4.5)

$$P_{2n+1}(x) = P_{2n-1}(x) + \frac{(2n)!}{2^{2n}(n!)^2} (1-x^2)^n x \qquad (A.4.6)$$

leading to the explicit solution

$$P_{2n+1}(x) = x \sum_{m=0}^{n} \frac{1}{2^{2m}} \binom{2m}{m} (1-x^2)^m \qquad (A.4.7)$$

Unfortunately, at the present time, no analytical solution for the corresponding equiripple polynomial exists.

A.5 MAXIMALLY FLAT ODD FUNCTION APPROXIMATING A CONSTANT

Consider the odd function $F_n(x)$ of degree n which is required to approximate to unity in a maximally flat manner around $x = 1$. We shall demonstrate that the solution may be expressed as

$$F_n(x) = \tanh(n \tanh^{-1} x) \qquad (A.5.1)$$

Differentiating, we have

$$\frac{d F_n(x)}{dx} = \frac{n[1 - F_n^2(x)]}{1 - x^2} = K \frac{(1-x^2)^{n-1}}{D^2(x)} \qquad (A.5.2)$$

where $D(x)$ is the denominator of $F_n(x)$ and showing that $F_n(x)$ is the optimum maximally flat solution.

A.6. EQUIRIPPLE ODD FUNCTION APPROXIMATING A CONSTANT

The odd function $F_n(x)$ required in this case must be optimally equiripple between the limits 1 and $m_0^{-\frac{1}{2}}$ for x in the range

$1 \leqslant x \leqslant m^{-1/2}$. Using techniques similar to those employed in Section 2.5, it may be shown that

$$F_n(x) = \text{sn}_0 \left(\frac{nK_0'}{K'} \text{sn}^{-1} x \right) \tag{A.6.1}$$

with the conditional requirement

$$\frac{nK_0'}{K'} = \frac{K_0}{K} \tag{A.6.2}$$

This follows from the fact that $F_n(x)$ must satisfy the differential equation

$$\frac{d F_n(x)}{dx} = C_n \frac{\sqrt{[1 - F_n^2(x)][1 - m_0 F_n^2(x)]}}{\sqrt{(1 - x^2)(1 - mx^2)}} \tag{A.6.3}$$

A.7 EQUIRIPPLE TWO-BAND ODD POLYNOMIAL APPROXIMATING ZERO

Occasionally it may be desirable to use an LC ladder network to produce a quasi band-pass response. In this case one requires an odd polynomial $P_{2n+1}(x)$ which is equiripple between ± 1 for $\lambda \leqslant |x| \leqslant 1$. The solution for this polynomial may be written in the parametric form[A.1,A.3]

$$P_{2n+1}(x) = \cosh \left\{ \left(n + \frac{1}{2}\right) \ln \left[\frac{H(M+u)}{H(M-u)}\right] \right\} \tag{A.7.1}$$

where

$$x = \frac{\text{sn } M \text{ cn } u}{\sqrt{\text{sn}^2 M - \text{sn}^2 u}} \tag{A.7.2}$$

$$M = \frac{-K}{2n+1} \tag{A.7.3}$$

$$\lambda = -\text{sn } M \tag{A.7.4}$$

using the standard notation for Jacobian elliptic functions and where $H(\theta)$ is a Jacobi eta function. Although a closed-form solution for the coefficients of $P_{2n+1}(x)$ is obtained directly, the zeros of $P_{2n+1}(x)$ may not be obtained in an analytical manner. These results may also be extended to the generalized rational function case[A.4] where the poles of the function are prescribed.

References

1.1 J. O. Scanlan and R. Levy, *Circuit Theory*, Vols. 1 and 2, Oliver and Boyd, Electronic and Electrical Eng. Texts, 1970.
1.2 A. Papoulis, *The Fourier Integral and its Applications*, McGraw-Hill, New York, 1962.
2.1 M. Abramowitz and I. A. Stegun (Eds.) *Handbook of Mathematical Functions with Formulas, Graphs, Mathematical Tables*, New York, Dover, 1964.
2.2 R. Saal, *Der Entwurf von Filtern mit Milfe des Kataloges Normurter Tiefpasse*, Telefunken, G.m.b.H., Backray/Wurtemburg, W. Germany, 1964.
2.3 T. Fujsawa, 'Realizability theorem for mid-series midshunt low-pass ladder networks without mutual induction' *IRE Trans. on Circuit Theory*, CT–2, 320–325 (Dec., 1955).
2.4 J. D. Rhodes, 'Explicit formulas for element values in elliptic function prototype networks', *IEEE Trans. on Circuit Theory*, CT–18, 264–276 (March, 1971).
2.5 D. C. Youla, 'A new theory of broadband matching', *IEEE Trans. on Circuit Theory*, CT–11, 30–50 (March, 1964).
2.6 E. L. Norton, 'Constant resistance networks with applications to filter groups', *Bell Sys. Tech. J.*, 16, 178–193 (April, 1937).
2.7 W. R. Bennett, 'Advanced problems no. 3880, 3881', *Amer. Math. Monthly*, 45, 389–390. (June–July, 1938).
2.8 E. Green, 'Synthesis of ladder networks to give Butterworth or Chebyshev response in the pass-band', *IEE Monograph*, No. 88 (January, 1954).
2.9 H. J. Orchard, 'Formulae for ladder filters', *Wireless Engr.*, 30 3–5 (January, 1953).
2.10 V. Belevitch, 'Tchebyshev filters and amplifier networks', *Wireless Engr.*, 29, 106–110 (April, 1952).
2.11 L. Weinberg, 'Explicit formulas for Tschebyshev and Butterworth ladder networks', *J. Applied Physics*, 28, 1155–1160 (October, 1957).
2.12 H. Takahasi, 'On the ladder type filter network with Tchebycheff response,' *J. Inst. Elec. Commun. Engrs. Japan*, 34, 65–74 (February, 1951).
2.13 L. Weinberg and P. Slepian, 'Takahasi's results on Tchebycheff and Butterworth ladder networks', *IRE Trans. on Circuit Theory*, CT–7, 88–101 (June, 1960).
2.14 I. Novat, *On the Element Values Problem of Chebycheff Filters with Arbitrary Patterns of Reflection Zeros*, EE Pub. No. 154, Technian Israel Institute of Technology, September, 1971.
2.15 S. B. Cohn 'Direct coupled resonator filters', *Proc. IRE*, 45, 187–196 (February, 1957).
3.1 W. E. Thompson, 'Delay networks having maximally flat frequency characterisitics', *Proc. IEE*, 96, 487–490 (November, 1949).
3.2 J. D. Rhodes, 'A low-pass prototype network for microwave linear phase

filters', *IEEE Trans. on Microwave Theory and Techniques*, MTT—18, 290—301 (June, 1970).

3.3 J. D. Rhodes, 'Filter with Periodic Phase Delay and Insertion Loss Ripple', *Proc IEE*, 119, 28—32 (January, 1972).

3.4 J. D. Rhodes, 'Filters Approximating Ideal Amplitude and Arbitrary Phase Characteristics', *IEEE Trans. on Circuit Theory*, CT—20, 120—124 (March, 1973).

3.5 P. H. Halphern, 'Solution of Flat Time Delay at Finite Frequencies', *IEEE Trans. on Circuit Theory*, CT—18, 241—246 (March, 1971).

3.6 J. D. Rhodes, 'Matched filter theory for Doppler invariant pulse compression' *IEEE Trans. on Circuit Theory*, CT—19, 53—59 (January, 1972).

4.1 J. D. Rhodes and M. Z. Ismail, 'Cascade synthesis of selective linear-phase filters', *IEEE Trans. on Circuit Theory*, CT—19, 183—189 (March, 1972).

5.1 G. L. Matthaei, L. Young and E. M. J. Jones, *Microwave Filters, Impedance-Matching Networks and Coupling Structures*, McGraw-Hill, New York, 1964.

5.2 J. D. Rhodes, 'Fourier coefficient design of stepped impedance transmission line networks', *International Journal of Circuit Theory and Applications*, 1, 363—371 (December, 1973).

5.3 N. I. Achiezer and M. Krein, 'Some questions in the theory of moments', *Translations of Mathematical Monographs*, American Mathematical Society (1962).

5.4 J. D. Rhodes, P. C. Marston and D. C. Youla, 'Explicit solution for the synthesis of two-variable transmission line networks', *IEEE Trans. on Circuit Theory*, CT—20, 504—511 (September, 1973).

6.1 T. A. Abele, 'Transmission line filters approximating a constrained delay in a maximally flat sense', *IEEE Trans. on Circuit Theory*, CT—14, 298—306 (September, 1967).

6.2 J. D. Rhodes, 'The design and synthesis of a class of microwave bandpass linear phase filters', *IEEE Trans. on Microwave Theory and Techniques*, MTT—17, 189—204 (April, 1969).

6.3 J. D. Rhodes and M. F. Fahmy, 'Proof of the recurrence formula for the distributed equidistant linear phase polynomials', *International Journal of Circuit Theory and Applications*, 2, 405—406 (December, 1974).

6.4 M. F. Fahmy and J. D. Rhodes, 'The equidistant linear phase polynomial for distributed and digital networks', *International Journal of Circuit Theory and Applications*, 2, 341—351 (December, 1974).

7.1 J. D. Rhodes, 'The theory of generalized interdigital networks', *IEEE Trans. on Circuit Theory*, CT—16, 280—288 (August, 1969).

7.2 J. O. Scanlan and J. D. Rhodes, 'Microwave networks with constant delay', *IEEE Trans. on Circuit Theory*, CT—14 290—297 (September, 1967).

7.3 H. W. Schussler, *Digitale System Fuer Signalverarbeitung*, Springer—Vierlag, 1973.

7.4 J. D. Rhodes, 'Generalized interdigital linear phase networks with optimum maximally flat amplitude characteristics', *IEEE Trans. on Circuit Theory*, CT—17, 399—408 (August, 1970).

7.5 M. F. Fahmy and J. D. Rhodes, 'Finite band approximation for distributed and digital selective linear phase transfer functions', *International Journal of Circuit Theory and Applications*, 3 (March, 1975).

8.1 A. Fettweis, 'Digital filter structures related to classical filter networks', *Archiv fuer Elektronik und Uebertragungstechnik*, 25, 79—89 (1971).

8.2 J. D. Rhodes and M. F. Fahmy, 'Digital filters with maximally flat amplitude and delay characteristics', *International Journal of Circuit Theory and Applications*, 2, 3—11 (March, 1974).

A.1 N. I. Achieser, *Theory of Approximation* (Translated by C. J. Hyman), New York, Unger, 1956.
A.2 J. D. Rhodes, 'Equiripple transfer functions with prescribed passband zeros', *Proc. 1974, European Conference on Circuit Theory and Design*, 223–228 (July, 1974).
A.3 R. Levy, 'Characteristics and element values of equally terminated Achieser–Zolotarer quasi-low pass filters', *IEEE Trans. on circuit Theory*, CT–18, 538–544 (September, 1971).
A.4 R. Levy, 'Generalized rational function approximation in finite intervals using Zolotarev functions', *IEEE Trans. on Microwave Theory and Techniques*, MTT–18, 1052–1064 (December, 1970).

Index

Algorithms   5, 135, 164
All-pass networks   46, 57, 77, 104, 141, 180
Analysis   1
Approximation
  amplitude   3, 12, 134, 206
  phase   10, 76, 171, 210
Arbitrary phase   90, 122, 126, 178
Arithmetrical symmetry   121
Auxiliary parameters   26, 34, 37, 141

Band-pass filters   29, 71, 120
Band-stop filters   29, 71
Bandwidth scaling   139
Bessel polynomial, see Linear phase polynomials
Bisection theorem   127
Bounded functions   132
Bounded real functions   3, 18, 114, 121, 126, 138
Broad-band matching   46, 141

Canonic   115
Capacitive discontinuities   149
Characteristic admittance   131
  admittance matrix   150
  impedance   36
Chebyshev functions   5, 8, 19, 22, 36, 40, 108, 125, 135, 207
  inverse   5, 19, 28, 38, 135
Circulators   102
Coaxial realizations   149
Complex coefficients   111, 122
Compression of signals   118
Continued fraction expansion   116
Contour integration   7
Cut-off frequency   8, 9

Definite integrals   78, 172
Delay group   5, 8, 11, 12, 107, 171
  phase   16, 88
Design tables   38, 132

Differential equations   24, 31
Digital filters   1, 206
Distributed filters   1, 134, 171, 184
Doppler effects   118
Dual filters   128

Ellipse   26
Elliptic functions   5, 8, 18, 30, 38, 50, 60, 108, 134, 217
  parameter   32, 217
Entire function   24, 78, 84, 88, 125
Equalization   106, 109, 184
Equidistant constant phase delay   88
Equidistant linear phase   83
Equiripple   5, 12
  optimum nature   13
Even functions   31, 216
Even polynomials   24, 103, 216
Exact linear phase   172
Explicit formulas   4, 19, 104, 166
  Chebyshev   40, 47, 139, 146, 207
  elliptic function   50, 59
  inverse Chebyshev   50, 59
  maximally flat   40, 47, 139, 146, 207
  reflection   101, 118, 180

Finite-band   13, 110, 121, 172, 178
Fourier coefficients   157
Frequency invariant reactance   38, 55, 70, 129
Frequency transformations   19, 29, 68

Gamma function   173
Generalized Chebyshev functions   149, 212
Generalized interdigital filters   193
Geometric symmetry   71, 121
Group delay, see Delay

Half power   21
High-pass ladder filters   29, 115

Hilbert transforms 7
Hurwitz factorization 28, 136
  polynomials 6, 103, 176
  sequence 82, 95, 125, 180

Imaginary resistors, see Frequency
    invariant reactance
Impedance input 1
  inverter 36, 115, 139, 143, 152
  scaling 19, 68
Insertion loss 64
Interdigital filters 149
Interpolation 76, 125

Jacobi polynomial 177
Jacobian elliptic functions 31, 60
  imaginary transformation 34
Junction reflection coefficient 169

Kuroda transformations 135

Ladder networks 6, 180
  synthesis 29, 35
Linear phase 5, 171
  phase polynomials 76, 105, 178,
      185, 210
  system 6
Logarithmic phase 104, 118
  phase polynomial 95, 118
Loss insertion 64
  return 64, 157

Matched response 118
Maximally flat amplitude 5, 8, 19, 36,
    106, 136, 157, 210
  constraints 110, 117
  linear phase 77, 110, 172
  logarithmic phase 95, 118
Minimum phase 5, 37, 170
Multipath 11, 127, 193

Non-minimum phase 5, 11, 105, 184
Non-reciprocal 102
N-wire line 135, 150

Odd function 31, 217
  mode admittance 115, 127
  polynomial 24, 103, 216, 218

Parameter shifting 175
Partial pole extraction 38
Periodic 134

Phase shifter 116
Piecewise linear 8
Positive definite matrix 162
  definite sequence 161
  function 39, 99, 114
  real function 1, 37
  real operators 163
  real residue 39, 82
Power series 9, 20, 139, 166
Prototype networks 38, 134

Quarter periods 32

Reactance function 87, 103, 115, 128
Reactance slope parameter 19, 71, 75,
    121
Reciprocal 2, 35, 104
Recurrence formula 25, 78, 168, 173
Redundant unit elements 165
Reflection coefficient 2, 102
Reflection filters 5, 101, 116, 182
Return loss 64, 157

Sampling 1
Scattering matrix 2, 102
Selective linear phase 11, 105, 184,
    210
Spectrum 3
Stable 6, 81
Stepped impedance filter 146, 152
Stubs 135, 153, 181
Surface acoustic wave 118
Synthesis 1, 37, 127, 193

Table 132
Transfer functions 3, 114
  matrix 103, 115, 143
Transformer 103, 129, 169
Transmission coefficient 2
  filters 5, 119, 127
  lines 1, 5, 136
  zeros 151

Unit elements 135, 152
Unitary condition 2
  function 45, 141

Waveguide 6
Wavelength 4
Wiener—Lee transform 7

Zeros 6